THE RELIGION OF PHYSICS

by

Philip C. Groce, MD

Copyright 2019 by Philip C. Groce

Also by the author:

When Mirrors Become Windows, Northwoods Press, Thomaston, Maine, 1985.

Personality and the Soul: Sixteen Women Show Us the Connection, Seven Coin Press, Rockland, Maine, 2001

Redefining Success: Working Close to Home, iUniverse, New York, 2005

ACKNOWLEDGEMENTS

I am indebted to the following people for their help in the preparation of this book:

My family for their love and support, especially my wife, and special thanks to Hazel for feedback.

My physicist brother, David, for his help on particle acceleration.

Dr. David Tyler for his help translating the manuscript from my 1980s discs.

Dr. William Geoghegan for direction.

Rabbi Harry Ski for insight on Judaism.

Gary Sukeforth for feedback and laughs.

Michael Keyes for digital design of the diagrams

Lynn Weaver for first chapter review

Brooke Nixon for a helpful referral

Deborah Mendelsohn for suggestions

Dr. Michael Bither for valuable conversations

Per Henrikson, for reading it 25 years ago

Dr. Allan (Chip) Teel for encouragement

PHILIP C. GROCE

Dedicated to my wife, Dianne

A truly kind person

CONTENTS

THE RELIGION OF PHYSICS	1
Acknowledgements	3
E	7
Elements of Healing	44
Defining Mind	55
Ownership	65
Plan	85
Light	111
Time	133
Personality	169
Being and Non-Being	190
Working with a Symbol	209
Transmutation	223
Psychological Mechanisms	245
Desire	266
Will and Space	279
Monsters	297
Purpose	311
Synthesis	325
Appendix: Light and Sound	376

Credits 399

E

Though in a few pages you will see the startling beginning of this project, the very beginning was in 1970 when I was sailing from San Diego, California to the Sea of Cortez in a 30' wooden sailboat—heavy winds, pitch-black night, high seas splashing into the cockpit. I had already struggled forward to set the storm jib, and I knew there was nothing more I could do. I sat gripping the tiller, and suddenly a great calmness came over me. I realized that I could do nothing more. The rest was totally out of my hands.

That instant, I was real. The most unimpressive part of it all was that I was almost 30 years old. About time. Until then, I didn't know what real was. Being in a psychiatry residency at the time (UCSD) likely aided this transition. Such an atmosphere catalyzes changes, they say. How many physicians go into psychiatry and solve their own problems? Maybe a lot.

I had joined the psychiatry residency on a whim while I was running a free clinic in San Diego in the Afro-American ghetto. Discharged from the Navy the year before with service in Viet Nam as a physician, a social worker and I started the clinic to provide services to families and also to individuals with drug problems. Unplanned by us, white hippie-types looking for treatment for their sexually transmitted diseases monopolized the services from the very first day. Every evening they lined up. What the newly discovered birth control pill wrought. Free love. And a free clinic to boot.

But later, the clinic began to also serve as the outpatient arm for one of the first methadone programs in the country, run by the psychiatry department at UCSD. The chairman of the department sort of talked me into going into the psychiatry residency. So, what? I thought, I didn't have anything else to do, and I could still run the clinic on the side. I could continue fundraising for the clinic at the intermissions of rock concerts, TV appearances, newspaper articles, meetings with politicians. World by the tail.

On return to the residency after my sailing break, the first thing I did after walking into the psych residency office was to inform the receptionist, a straightforward Afro-American lady, that I was now REAL. "Welcome," she said with her beautiful smile, extending her hand. I felt like a junior member of the human race, a race I thought I was winning. Winning? What am I doing here? Alice in Wonderland.

Right then I decided to leave Wonderland—California-- the place of my birth and find somewhere to settle, hopefully married, raise a family, and practice my brand of medicine in a small community. I couldn't do it in California, participating in what I called the California Neurosis, every day being beautiful and possibilities infinite, feeding my own arrogance (though now, 50 years later, California undeservedly suffers in the throes of climate change and natural disaster).

I wanted a defined community with a sense of history and not much growth potential where I could write, though I could not precisely define my vision. I thought I had something to say, not knowing how elementary my thinking was. But it WAS a start.

Resigned from the residency after the first year, found another medical director for the free clinic, packed up a VW van converted to a camper, and set off to find a place to live.

After exploring through Northern California, Oregon, Washington, and then across Canada, I ended up on an island in Maine. I worked for a year in a small medical clinic learning how my California ways were definitely out of sync with Maine life.

I made some changes, and after a year I opened my own family practice in a farming community in Maine. After dating for 5 years, I married a local girl, and we made (and continue to make) a life together; and we remain in the same old house that requires constant attention—it's like a living organism, or in this case, one on prolonged life support. We have raised four kids into adulthood, though we lost one child (our fifth) through an accident when she was one year old. We are a multi-racial family with all the kids adopted as babies.

My wife and I ran the practice, she doing all the office work, and I saw patients in a converted dining room, the patients waiting in the parlor. Easier to raise the kids together, having the office in the home. Had office hours and made thousands of house calls. Back then, the office call was $7.00 and house call $12.00. I still work on a limited basis in my field, geriatrics, though we no longer have a practice in our home. We love the old people. They tell it as it is.

I continued to write, and my first book was novelettish, *When Mirrors Become Windows*, motivated by my own experience in the old boat and showing how resolution of neurosis worked. Continuing to write, I usually started at 2 am and worked to 4 am, then back to bed for a couple of hours, thence to daily living.

One night over 30 years ago I was typing away at a rather self-centered novel, and I paused to consider what to write next. The main character, a politician, stood in front of an audience, and someone had asked him if he was religious. He stood there thinking. I pondered along with the character, pondering . . . and I automatically typed, "God is energy." Instantly, I experienced what (I later learned) is called "The Raincloud of Knowable Things." That Raincloud so overwhelmed me that I cannot describe it other than to say that an entirely new operating system flooded my mind. I grabbed a blank tablet and began scribbling down as fast as I could the material coming in. I wrote for hours before the Raincloud subsided. This was not drug induced.

Exhausted, I examined what I had just written. I had no idea where it had come from, and not only that, I had no idea what the material was. Philosophy? Religion? Physics? If so, what philosophy? What religion? I searched to find what to call this material.

After some thrashing about, I had direction from a professor of philosophy and religion at Bowdoin College in Maine, William Geoghegan. I began reading philosophy, history, history of philosophy, religion, physics, and anthropology. After 10 years of study, I felt that I had a grip on what it was about. I also studied 7 years with the School for Esoteric Studies. Outside of all that, I read an entire set of Encyclopedia Americana and also the many volumes of The History of Civilization by Will Durant, a man I greatly respect. That reading helped put my thinking into context.

When writing I learned to center myself in the same frame of mind as when I had the original Raincloud experience. That means I had no way of knowing the direction I would be going, but I dutifully wrote what was coming to me. It took years. With this method of writing, each chapter builds on the preceding and prepares for the next. It's a space-age perspective that can be applied to any belief system (including atheism) and thereby enhance that system. It mostly fills the void in the minds of many who look at the world in scientific ways, not really accepting religion, and generally landing in agnostic territory. Still, even agnostics yearn for something ultimate, which as we know, is the subject of religion.

I wrote about things I never expected, and ultimately ending where I never thought I was going. Where I went is to the very instant of change, and this book, though it does not seem heading in that direction at first, certainly ends there. Some things appear so fantastic that I have to say that I do not understand them completely, and when I do not, I tell you. Still, I found plenty to laugh at, and hopefully we will have some laughs together along the way. This material is as serious as serious is serious, but, still, it does not hurt to have a laugh now and then. I am not trying to fool anyone, nor am I planning to make money on this endeavor. It is just that I promised myself I would get this book out before I died. At the moment, I'm still living.

You should not accept anything I write unless it rings true to your inner self. I am rarely footnoting or sourcing. You will know if it rings true or not. That is the only verification that counts. This first chapter is the bottom line. The rest is infrastructure. The infrastructure contains the physics and geometry, but don't recoil from that. It's mostly high school level stuff with some basic quantum mechanics thrown in for those who have interest. But if you don't like the bottom line in this first chapter, then no need to go further.

The rest of this chapter after this introduction will contain mostly material from that first experience, including more on the its mechanics. This chapter covers lots of ground, as did that Raincloud, but it sets the stage by bringing up subjects for the rest of the chapters. One thing builds to another, and eventually some astounding (at least to me) conclusions are reached, though I do not think they can be understood unless the reader begins at the beginning.

All the material remains as written originally, though I have updated examples and some aspects of quantum mechanics. After I completed the original manuscript in the 1990's,

I set it aside thinking that no one in that day and age would want to read it, let alone want to understand it. I decided to wait at least 25 years to publish it. That time is up, and I'm almost 80. If there ever is going to be a time, this is it.

What is E?

We cannot go beyond our laws of physics. These laws restrain us and determine what we can do. For instance, we cannot counteract gravity, nor can we humans attain the speed of light. We can take a step further and ask: What *are* the real laws of the universe, the laws or boundaries that allow our *known* physical laws to work? In other words, our physical laws are only *effects*. Effects of what? What are the *causes* behind our known physical laws? That is physics beyond physics, or metaphysics—the supposed framework behind, and holding together, and causing our physical laws. All metaphysics is conjecture, or it would be physics.

Energy is everywhere. We know that matter and energy can interchange. That important fact we realize from Einstein's famous equation. According to Einstein $E=Mc^2$, meaning that Energy (E) equals Mass (M) multiplied by the *square* of the speed of light (c^2). Light is fast, but the square of the speed of light is so fast that it is beyond conceptualization. If we try to accelerate particles to the speed of light, they need more and more energy the closer they get to the speed of light. Physicists say that it would take an infinite amount of energy for any particle to attain the speed of light. Good reason to slow down on the freeway. The entire point of this book is to say that all is energy. Matter and energy can interconvert. That perspective is E, and closer to physics than anything else. Allow me to explain.

Let us say we have a box large enough to contain the universe, a universe that appears endless, but maybe it isn't. We might be connected to other universes by various means

(Multi-verses), perhaps involving quantum entanglements. All that is based on speculation emanating from solid physical research by some of the smartest people in science. That speculation, until proven, touches the realm of metaphysics. I will not be dealing with metaphysics, but I am using the concept of a defined universe, calling it "within a box," only to say that I am not considering anything beyond it, if there is anything beyond it. E, that we know, is in the box, though E in-itself could possibly extend beyond the box. But that needs considerable further discussion.

Except for light (which contains so-called photons), all energy that we know is attached to some physical (material) system in some way, or we would not know about it. We do not know of energy unassociated with matter. I am looking at the universe in terms of Einstein's equation which equates energy with the relationship of light and matter. We know of no other E (Energy) not associated with matter. If there existed such energy, we would have to call it Energy-in-Itself, energy unassociated with matter, and that is beyond relativity and Einstein's equation.

Religious systems base themselves on a connection of the so-called spiritual world to the material world. The movement from a life *less* spiritual--being more materially oriented--to a life *more* spiritual and less materially oriented--equates to what religions call enlightenment. But there remains a step beyond this direct connection to the material world when religions define their concept of God. That step would be Energy (their Spiritual) not connected to matter. Deity is then defined as pure Energy, or Energy-in-itself, called Father in Western religion, or The Absolute in Eastern religion, or Absolute Spirit in esoteric terms. I cannot fault that step since religions base themselves on that faith. No problem. But I do not make that step, and I will not ask you to use faith in this book on energy. If you *want* to make that step, then E *can* be defined as Energy-in-itself, and it does work, but you have to want to make that jump. It's up to you.

Is a universe conceived in a box ultimate enough for a religion? Am I insisting that you be a materialist? You will see that I am not doing that, though again, E can support a materialist outlook. I believe I can show that it *is* workably ultimate and irretrievably connected to the ethics which all religions purport. Not only that, I can state definitely that an acceptance of the importance of energy as a concept can add depth to a person's beliefs even if she/he stays firmly within a particular religious system, no matter how much cultural dogma is also attached.

This entire book shows how a belief in the simple statement, God is Energy, plays out in daily life and how it can be deconstructed into its components and astonishingly reconstructed into something well within established and accepted ethical norms. It is a trip worth taking--for some--but not for all.

So, how do the cards of life unfold with E? The rest of this chapter I will show E's relationship to religion, economics, a new species, art, kindness, and celebration. The chapters after that explain the infrastructure of looking upon God as energy, but thankfully, each chapter covers one particular subject, making understanding a bit easier.

I do not propose to convert you to anything. I do not expect you to accept anything I write unless you sense it within yourself. And I say one last time that E is merely the perspective that God is energy, which includes any way you want to characterize God in your own perspective or established religion. All that need not change. If you want, E can also stand on its own legs as a religion—if it appeals to you. No faith required.

And please remember that this chapter just introduces subjects that I will later discuss in much greater detail. Nothing in this chapter is explained completely, and you will have questions. Using this chapter as an end in itself is a mistake. Read it, think about it, maybe read it again. If it has interest to you, you need to read on and allow yourself to form larger conceptualizations *within your own system* of thinking. I have done the best I can to make it understandable. This first chapter shows the bottom line, and the following chapters tell how it all comes about and where it comes from, and where it ultimately goes.

Not easy material. It is like explaining art or poetry, and I believe that is the biggest criticism of this work: Why not just leave it to art and poetry? I cannot argue with that, except that this is my art, and I know there are some people, not many, who want to know the inner workings of what I am talking about. The inner workings are called by the term esoteric, and that is a big subject; and I am not adept in that field. I will use esoteric terms and conceptions at times, when appropriate. Mine is a different approach with the same end, but I could not have written this without background in esoteric studies, and I thank them for that.

E and Religion

We house religions in institutions of religion where people congregate, almost always with priestly hierarchies. According to the perspective of E, institutions of religion suffer from the inflexibilities of human error that all institutions suffer from—just as with schools or government. Institutions function politically, and they naturally have interest in finance and power.

Regardless what an institutional religion purports, the institution itself has its own problems, and it tends to evolve into a barrier between the person and her/his conception of God, however conceived. Organized religion can institutionalize discrimination in the guise of what it considers right or wrong.

I hate to say it, but people who have claimed religion or

a 'higher' mission of some sort have caused the greatest suffering in this world. These are hypocrites, people who hide behind the cloud of a belief system and carry out carnage 'for the greater good.'

The ultimate form of hypocrisy is sexism fueled by religion with suppression of women. I can make a good case that the major conflicts in the world are caused by the ultimate hypocrisy: Men should run the world.

We have been taught to watch out for the psychopaths in the world. Psychopaths (or more modernly called sociopaths) are people without conscience. But hypocrites, even with their consciences, pose greater dangers than the psychopaths, and they greatly outnumber them. Hypocrites seek power and lead groups. Having people follow them reaffirms their own twisted thinking. The hypocrite *has* a conscience, but guilt looms because the person cannot do what he/she wants to do and still stay within the boundaries of the religion or belief system, or political system. Who wants to feel guilty? One must either suffer and feel guilt, or else erect moral blinders to block it out. Hypocrites replace guilt with 'good intentions.' Many people live on good intentions. We all have seen that.

In contrast, the psychopath doesn't need to erect moral blinders since she/he has no interest in morals or any other conceptual frameworks that might hinder getting what is wanted. Fortunately, the nature of psychopaths usually prevents them from being able to stay organized long enough to attain much political power, though the history of humans remains rife with hypocritical politicians.

Every day we humans have many choices as to how to act. Humans can act in only a few ways and live in peace as opposed to the many ways we can cause disruption, chaos, and war. The very basis of the Ten Commandments and the Golden Rule is good sense. Greed and pride necessitate the moral blinders of the hypocrite. On the other hand, organized religions can serve a significant cultural value by aiding people who feel that they cannot in themselves control their wayward ways, not to mention the community aspect of a congregation. There is value in all that. For many, organized religion is the perfect solution.

Some religions want to convert the follower. If conversion is successful, the convert then sees reality through the perspective of that religion. 'Born again' is a good example, though not the only one. Unfortunately, the convert will think that her/his new perspective is the *only* truth, that anything else is falsehood. The new perspective seems perfect, and it is. But it is here, and only here, that E separates itself from religion. E does not accept exclusivity.

Allow me to explain what I mean by exclusivity. All religions have equal perfection as conceptual systems. The fundamentals of all religions are the same. Placed in cultural references, they appear different. They all equally have imperfection in the problems associated with their institutionalization (Some will deny that, but I do not believe them.). E would accept all religions, along with conversion, but E would maintain that no religion can say it is the *only* path. Hence, E cannot accept exclusivity.

The only truth is that there is no only truth. That's a bummer, I know, but there is nothing relative about energy in itself. Immediacy and relativity take up big sections of the book, and we are heading to some conclusions that will surprise you, especially regarding time dimensions. Think about it, and if you refuse to remain open to something different, you might as well go no further... but I think you are going to miss something important. You need to see how absolute and relative can both exist and live comfortably together, really; but we have a long way to go before I can show that. I am talking about the big picture. Let me explain.

Remember at the beginning of the book I mentioned the free clinic in San Diego where I was working as the medical director? Before that, I was traveling, actually, bumming around the world. When I returned, I went into the residency to become a psychiatrist. One older member of the psychiatry staff served as advisor to the residents. We each had periodic conferences with the advisor during the year. Marvin Karno was his name. Marv had been around the barn and back again, and I have always remembered something he said to me.

During one conference, Marv leaned over to me and said in a quiet voice (as if there were other people in the room, which there weren't), "You know, Phil, you are the most un-psychopathic person I have ever met." I had to smile at that, because that is about the biggest compliment a psychiatrist can give (But you have to think like a psychiatrist to realize that).

Then he said, "Most people live in very small orbits, but you have the largest orbit I have seen." All of that was nice of Marv to say, and I thank him. I learned something about myself I had not realized: I was an un-psychopathic orbiteer. That describes me, but I don't think I would include that on a resume'. But I am telling you this, joking aside, to warn you that my perspective *is* large, and you may not see reality in that way, and that's okay. It is just the way I am put together. There are some odd ramifications of that, and one of them is that I can interpret dreams. If someone tells me a dream, I can see it all come together in a meaningful way very quickly.

All of this may seem a bit off the subject, but you don't know me from nothing. I have been busy living life, working, and writing. And I do not intend to go out and sell the world on this book. I am just putting it out. Maybe a few people will read and understand it, and if they find it of value—EVEN ONE PERSON—then it has been a success. After having published 3 books, I decided not to go through the usual publishing process again, and that is why I am putting it out myself through Kindle. I did the copy editing, but all my hand-drawn diagrams have been converted to digital by a true professional. You can find minor errors, especially in formatting, and I am still working on changing that, as this type of publishing allows changes. It is readable, but not always easy. Please forgive the imperfection.

Back to the business at hand: I am now going to discuss aspects of religion that unnecessarily conflict with E.

E and Intelligence

The intelligence behind the creation of the universe remains an important subject in religion, called Universal Intelligence or Intelligent Design. In my view, there are basically 3 ways to look at Universal Intelligence.

First, there is the Christian, Judaic, and Islamic concept of intelligence which I call 'Western,' and it maintains that intelligence in the universe exists, and we can understand it through what we see in His works and through revelation and through what are believed to be holy books. They all use the pronoun Him to refer to God thereby making this God a personal God. What can E say about this conception of universal intelligence?

E can say that through analogy we can learn about the nature of the universe. We are surrounded by analogy since the physical laws of the universe only allow us to create within those boundaries. Everything reflects those boundaries. Everything that we know is within those boundaries. Hence everything created within those boundaries is analogous (Wow, what a stretch!). Both nature and humans cannot create anything outside of physical laws. Everything within those boundaries reflects the greater picture of reality. Such is analogy, and it *is* the big picture. How could it be otherwise? Please tell me.

The physical laws that frame the universe force everything that takes place in the universe to be analogous. I could say, "Air is made of tomato juice." I am capable of saying that as I am capable of saying anything, because our physical laws allow me to do that. I am allowed to make mistakes, and there can be chaos. If the universe is intelligent, then it is intelligent enough to allow what we call chaos and mistakes. Chaos is just a term we use when we cannot envision or prove a bigger picture (Some physicist would disagree). If we could see the BIG picture, we might see the contribution (of some sort) from chaos or mistakes on physical systems that we understand. Mistakes, after all, set the groundwork or the *beginning* for learning and change. The big picture, the frame of it all, remains the same, and there just might be other universes (multiverses) interfering or linked to us in very odd ways. No one knows. I will be defining chaos in relation to E much later on, an important part as it turns out.

Another way Universal Intelligence is viewed I will call the 'Eastern' concept of Universal Intelligence, and it is considerably simpler. I will use the example of Buddhism. The tenants of Hinduism are basically Buddhist, though in a complicated cultural setting.

Buddhists maintain that intelligence in the universe does exist, but we cannot understand it. This is the concept of the Absolute--too absolute even to be named, much like an Entity beyond energy as we know it, attached to matter. Who could argue with that? Energy-in-itself? What is energy, by the way? Our words fail, our intelligence fails. We have to give up. Unless, of course, we use analogy and look inside the box

Finally, there exists the conceptual framework that says there is *no* underlying intelligence in the universe whatsoever. This we call atheism. E can admit to a natural energy flow toward the most efficient. Behavior is guided by good judgment and laws. Energy exists. Nothing negative about that for an atheist.

What might science say about the existence of Universal Intelligence?

Science says that which survives on this earth does so because it utilizes energy in a more efficient manner than its ancestors. The more efficient will survive in the changing conditions of our earth. We humans find ourselves somehow placed between the heavens and earth while having the capacity to change things by our own choice. E supports the path toward the more efficient, which in human terms, is the path of creativity, and more importantly, the path of the energy called love (as I will explain later).

Within the perspective of science, the intelligence behind E is expressed by the more fit ultimately surviving, which science refers to as natural selection. E would refer to natural selection as the process of survival of the more efficient. *Energy drives that system.* In the most elemental conception of the beginnings of life on earth, scientists say that organic molecules came together through attraction forming larger units that react to one another thereby forming the beginnings of life. How could we call that chance, since the molecules are attracted to each other through physical laws already set; and they react in only certain ways according to the laws of physics?

And *where* did the physical laws, the very boundaries of our lives, come from? Through energy. That energy is defined by E as the speed of light squared multiplied by the mass of the universe. We partake of that energy in small part when, for example, we create, even when we screw up. Such is intelligence

I will now extend the discussion to include artificial intelligence, since that is now considered a type of intelligence.

Science serves as the discoverer of universal secrets.

Analogy surrounds us. For instance, the network of our brains reflects networks everywhere. The inventions we make through our creative energies reflect what is possible for us to conceive within the universal bounds of energy. This is analogy at work, just as the computer echoes the inner connections of our own brains, reflecting the social networking enabled now through the computer. For example, allow me now to go back to the perspective-changing experience I described in the beginning of this chapter, the Raincloud of Knowable Things.

Reprogramming the brain is much like reprogramming the computer. The brain, however, has the capacity to learn on its own through experience and combine that learning by energizing circuits related to these memories. This learning is not represented by circuits in just one place. It is scattered in different places, depending on how the learning originally occurred, the setting, the emotional tone, and many other factors. Learning continues to add to the volume of energized circuits, and these circuits have connection with each other, but tenuous, in as there is not yet conscious connection between them. We do not *realize* the connections between them.

When the critical mass of learning, the setting, and the temperament is right, then and only then, do all of these bits of learning of a sudden come together represented by a conclusion, a conscious insight. Ah ha! The result can be an idea, big or small. The more circuits involved, the grander or more inclusive the insight. With insights encompassing universal conceptions, then the conscious brain has undergone a form of reprogramming much like a computer goes through. In humans, this process can also be called synthesis, with the trigger of the synthesis being motive. Motive is everything. Synthesis and motive will be taken up in great detail later in the book.

The difference between the computer and our brains is

that the computer is a machine, and it depends on us programming it. In our brains, the process just occurs. The mechanism of evolution of learning (including the so-called evolution of spirit) will always separate the computer from the human brain. In essence, this entire book examines that difference, despite deep learning and the sought-after master algorithm applied to AI. More on AI later. Also, more on the Raincloud experience, as there is much more to it than what I have described here.

Concluding this section on intelligence, I have said that E is in the universal box and hence does not compete with any religious conception of God, even out of the universal box. E just says that energy exists, and the more efficient is that which survives within the box. The energy of the entire system is characterized by the speed of light squared multiplied by the mass of the entire universe.

E can accept all of the conceptions of universal intelligence, West, East, atheistic, and science. Considering that such is possible, then how does each of us USE our own intelligence?

How We Use Our Intelligence

We each learn through experience, even if that experience is reading a book. The workings of our intellect, including logic, work okay in our everyday lives, but they fail in matters of importance. In important matters, intuition and revelation are quicker, more inclusive, and more reliable. Intuition and revelation remain the basis for creativity and the cornerstone of higher intelligence and putting intelligence to work, adding to all which a person has learned before.

Both fortunately and unfortunately, we humans attach emotion to nearly all our thought processes. When we have any experience, we also have an emotional component to that experience. Our thoughts become automatically attached to an emotion. Emotion brings us back to earthly desires, but those desires can blind us from seeing beyond ourselves to higher levels of thought. I will discuss intellect and emotion much more in the body of this book, especially in the chapter on Mind.

When thinking about how humans make decisions, we have to admit that we cannot look at any situation without emotional and intellectual attachment to our experiences from the past. This attachment to the past allows continuity and application of past learning to the present situation, but it can also blind us. This blindness we call prejudice, basically, pre-judging. We all lose the unprejudiced sense of discovery we had as a child. To protect ourselves from (feared) emotional hurt, we erect barriers as we grow older. Those barriers remain the biggest obstacles within each of us in our personal growth. Barriers of pre-judgment also limit science in a big way in its quest to discover the secrets of the universe. After all, the universe surrounds us *all* the time, its secrets sit there, merely waiting to be discovered, though veiled from us because of our ignorance.

I am not saying that emotion is evil, and we should all be intellectual automatons. I am only saying that the so-called negative emotions such as fear and anger can attach to thought. Emotions attach to thought and introduce possible error and serve as obstacles to further learning. Emotions do attach to all our thoughts in greater or lesser degree. Such is life; such is being a human. I am glad of it. Where would novels be without it?

At this point I caution you to maybe take a break, maybe come back to this material at another hour, or maybe re-read what you have already read. The fault of this entire book is that it is chock full of material (especially with this introductory chapter), and if you try to take too much in at one time, it can fill up the mind and cause a clog and prevent further reading from having any meaning. I get that feeling, too, when I proof-read my own material, and I'm the one who wrote it. I ought to know better than write something like this. An editor of one of my previous books once said to me, "Why don't you write a love story?" Of course, I rejected that. My loss, I suppose, though love plays a critical part in this book.

How can anyone learn to drop the emotional tangle of desire that attaches itself to thought thereby blocking access to higher energies within us, those of creativity, revelation, and genius? We see different techniques: religious or philosophical study, meditation, long quiet vacations, drugs, prayer, dream analysis, challenging nature, plus many others. If successful, even in a small way, we feel we have brought ourselves closer to something more *inclusive* than just ourselves.

Some people want something more personal than just science and self-experiment, and they seek a personal God, one they can call Him (or Her?). All new revelations about ourselves or our surroundings have peculiarity to ourselves. The results of that creative effort nearly always expose us to new horizons unplanned by us. Looking ahead, it's nothing but black. Having a mind full of pre-judgments from past experience or clinging to the emotions of fear or desire results in a foggy perception of reality. Looking back, it all makes sense. Logic works well looking back, because it all fits together now that we can see in retrospect all the parts that we didn't realize before. Using logic to predict the future, however, is a mistake. People are too complicated, as is Nature. Much more about that later—in diagrams.

In summary of how we use our intelligence, I have tried to point out the trip-wire of emotion in new learning. Emotion is the drama in life, but it can hinder higher learning through pre-judging, also called prejudice. It's what love stories are all about.

E and Determination

Free will is a big deal in religion, and it is beginning to be a big deal even in science. How does E stack up with religious conceptions of free will? I will just quickly list some of the important possibilities for free will. It can get pretty boring if I go beyond that.

1. Our acts are predetermined, but we should act correctly because it makes sense.

2. Our acts are determined by fate. We should do what we want.

3. There is no determination. If we do the right things, we will be rewarded.

4. Our acts are not determined. There is no God.

5. Our acts are not determined, but there are choices (paths.)

The perspective of E maintains that all concepts of determination have truth, but determination itself is moot. Again, we deal with the fact that everything that we know about in the universe, so far, is analogous. So, the workings within each of us is analogous to universal workings. Energy drives the entire system.

We have two choices. First choice is to take someone else's word for what we should do and how we do it. I cite holy texts. From holy texts we can derive pretty much any conclusion regarding determination or free will. That is fine. All holy texts still contain considerable wisdom.

The second choice is to figure out what we want to do by looking within our own selves, as our own selves have analogy to the intelligence or the physics of the energy of the entire universe as we know it (I will be discussing the concept of self in detail in later chapters.). We can do what others tell us, or we can figure out things for ourselves. If we are going to figure out things for ourselves, then we must know how to be true to our inner-selves.

To *BE*, rather to than *react to* given (religious) laws, necessitates no restraints. If we act to adhere to laws, then we will eventually fail and succumb to hypocrisy. To BE, rather than react *to*, gets rid of 'supposed to,' and the laws that all religions have in common are automatically adhered to without a thought. It's automatic, really. BEING is what makes determination moot since any actions that result from BEING are congruent with the flow of energy. But it's a thin blade one walks on. So how can a person BE rather than ACT?

We journey to BEING through a creative process. Resolving our adolescence remains the first step, and many people will not take that step until later in life, many people, never. Resolving adolescence makes us REAL. You know it when you are real. Others know it, too. If you do not know what it is, then you are not real.

Resolution of adolescence involves taking a chance on yourself, being yourself, having the courage to be yourself. Only after we resolve adolescence can we attain additional changes in reality perception through an endless series of large and small creative experiences. With large changes, we enter a new paradigm of life. The process continues until we die, at which time our physical bodies turn into energy via the microbes. Embalming only delays that process.

In summary of E and determination, I will say that we can bog ourselves down thinking about free will and determination. E goes beyond those concepts with the concept of Being. I will discuss Being in much greater detail later in the book as well as the important relationship between Being and consciousness. At this point I should not portray Being as allowing perfection. We all are still humans.

Since I mentioned death (above), allow me to ask the question: Is it possible to describe other energies within us, after death? How would E deal those matters, especially since death and after-life are important subjects in religion?

E and Afterlife

Without going into detail, which I will do later, how does E stack up when considering what might happen after death?

When a body dies, it ceases to utilize energy on its own as a living organism. The coherent energy of our life merges into a more inclusive body of energy, and we become one with that. Such is the concept of heaven. E can accept that.

Energy is never lost, and that energy continues in other lives. Since that energy could go on to other human lives, E could accept reincarnation.

E also accepts NOTHING after death. The energy that made us individuals still exists after we die, which is no-thing, also. An atheist cannot deny energy. Energy does not disappear.

No matter how one looks at afterlife or no afterlife, E can go along with that idea comfortably. It is not a big deal, nor anything worth talking about or getting upset about. Institutions of religion like to talk about those things, but that is their business model. It sounds crass and disrespectful, but institutions like churches *are* businesses (I do not exclude the so-called Spiritual industries.). If their only model was the Golden Rule, there would be no problems. But churches, as with governments and businesses if they have been in business long enough, they develop complicated policies, doctrines, and hierarchies. They argue about these complications. After a very long while, the complications are what keep them in business—keeping up and justifying the complications. That is hell. It also leads to war.

Within the concept of hell, E has no function, but each of us can have hell, every day, right in our own lives. The religious concept of hell instills fear and guilt, the tools that motivate people to stay in line. If we can BE rather than ACT, we need not feel threatened to behave. To BE automatically moves us in the direction beyond the good. No need to fear. We are our own enemies. We each create our own hells. How BEING can be attained, rather than ACTING, takes up big sections in the book.

Any act *other than* one purely motivated toward the flow of energy has both good and ill-starred consequences. We label the most unfortunate consequences evil--we could just as well call it insensitivity, but the word 'evil' has had its way. The waters are calm before we dip in our fingers and produce ripples, then waves, and then reflected waves coming back at us. The waves we make in our lives produce both good and evil. We create the evil. Evil acts move us away from the flow of energy. We do not intend the evil nor do we want it, but we can always justify it as the only way, the necessary byproduct to bring about good as we see it. We pave the road to evil with good intentions--a toll road as it turns out. Later, I will show that all acts are on a spectrum of the energy of love, and the least of it, a perversion of love which we call evil, and the greatest of it we call holy.

To conclude this section, the perspective of E can accept any concept of life or no life after death. It also maintains that to BE actually, and automatically, results in non-intentional behaviors which all religions root for. I can hear the cheering now.

I will next show how E relates to economics within government.

Break time again? Maybe not. You may be a lot smarter than I am.

Economy and a New Species

I first must define economics in terms of energy. Money is concretized energy. You can purchase anything material with money. Money represents that power. As such, economics is the physical (or material) manifestation of energy. It is actually the *physical* circulatory system any country. To better understand economics, I asked my daughter, Hazel (then 8 years old) to draw us a picture. Figure (1) Tree of Homo sapiens.

The tree is basking in a nice summer day, and a raincloud is doing its thing (this is Maine, after all). If we say the tree represents the economy of our country, then we can see the roots, the trunk, and its extensions into branches and leaves to represent the economic system. Within the economy, the branches could represent the divisions of the economy such as business, education, science, religion, military, and government.

Out at the ends of the branches we find the leaves and flowers where we see the process of photosynthesis and hence, creativity. Light from the sun and raw materials create new life. The seed dropping to the ground represents the new life created. Considering the evolution of future trees, we know that many years from now the tree will look different and act differently. The more efficient and more adaptive trees will have survived.

Figure (1) Tree of Homo sapiens

In economic terms, what will be the most adaptive aspects of an economy that will allow it to survive and thrive? If a country cannot support itself economically, then it likely will not remain a country unto itself very long. If we look at old maps, we see that countries come and go. We cannot blithely assume any country is going to have its same borders in the future—including ours. People vote with their feet, and no one can predict the mass migrations of people. Conquerors can draw lines on the map and make countries, but as astronauts point out: There are no lines seen from outer space.

Yet every country thinks of itself as exceptional. If the leaders can convince the people that their country *is* exceptional, then it's much easier to ask the people to die for whatever the leaders want. That works in our country, too. Try to convince people in any country that they are *not* exceptional, and they will pull out their flag--and hopefully not their gun. But you would have to agree with them, in looking back, that the origin of their country *was* exceptional, at least to them.

So, we can look on the life blood of our country to be its economy. It is like the tree and its branches, and the seeds that fall to the ground being the future ideas concerning how the economic system will evolve. Nothing is static. We know that there exist in the world different economic systems competing all the time for dominance or success. Will any economy help accelerate the evolution of the people to be more adaptable and peaceful? Let us look at that.

As a species, we are no different than that tree. We all are one species, Homo sapiens. That means we are knowing and wise (sapiens). Certainly, compared to other animals, we are. But we're *not* smart enough to stop killing each other when we disagree. How smart is that?

How much longer can all this go on? If we say that someday we will all live in relative peace and appreciate everyone's differences and accept them and evolve together, how long would that take? At the present rate, a long time. Cultures conflict. Wars ensue. What is it going to take for us to progress beyond that, to a peaceful world? You want a common culture? Let's kill some more.

In the past, Homo sapiens was not the only species with human-like characteristics. Our species is the only surviving one, but that doesn't mean it will be the last. Our species will change, economic systems will change, borders will change. Everything will change.

Change is inherent in all species. Since that is a truism, we must accept that a new human species has already started among us (just as there are variations already existing for that tree). This new human would eventually be much more adaptable than our species today. We cannot predict from what quarters the seed(s) of a new species will evolve. Will it come from the rich and powerful? Will it come from the children of misfortune? We do not know.

What we may know is that the most immediate change in our species will be longer life. If we learn anything from experience, then we will learn even more because we all will be living longer. A big change in life expectancy should occur in the next 100 years. Wisdom, representing the good and more adaptable things we learn while we live, will take on a higher dimension for no other reason than we will live longer. What other features might go into making a more adaptable human species? Genetic manipulation through science may play a part. But what is the relationship between the evolution of that new species and governmental economic systems?

As I have mentioned, that which survives in the long run is the more efficient. The energy that drives the evolutionary process toward the more efficient, is what I am calling E. That energy has to be termed the ultimate good, beyond good actually, since it is the *only* energy we have. Religions refer to God as love. Therefore, going along with religion, we can term the flow of energy as Love. Spirit (defined a bit later) is also a word that refers to energy as we experience it. Therefore, the path toward the more efficient is the path of love, and creativity, as it turns out.

Only in recent history have humans had the ability to change their own environment in such drastic and powerful ways. We have big choices. How do we make the right choices, choices in the direction of the most efficient? Since destruction of ourselves as a species is so much at hand, the changes for a new survivable species will not be so much a change in physical appearance (I would exclude size, since it can depend on nutrition) as change within, and how the species adapts. Such change will not occur over thousands of years as occurred before, but much faster, as everything is going faster nowadays through technology. The adaptations, made internally, will survive as the more efficient over generations. Those changes would become inherent after time. It is not only physical appearance that has inheritance. Behavior also has inheritance. This is beyond culture. Postmodernists might cringe at that thought.

As it turns out, the energy of love is the most efficient, and that applies to economics just as well. Now we are back to the economy. The energy defined as Love is the energy of evolution, and economy is the circulation of energy within the culture. Love also is the energy of creativity and therefore the engine for innovation. Consequently, those who compete only to win, will lose. And if life is a game, then winning *is* everything. It may seem like it at times, but life is *not* a game. Motive becomes the most important thing of all.

A good action is more than good, even beyond good or evil. We have no word in English to denote that other than holy. These are acts guided by the energy of love, love being the manifestation of Spirit. **I will define Spirit as that realm within the speed of light squared**. More about that later.

Some authors talk of self-realizing acts, or purely self-conscious acts, but I will use the term CREATIVE ACT. The energy of such a creative act in each of us derives from our own individual creative centers (creative centers will be discussed later). Since everything is analogous, we can say the same thing about economics.

Whatever economy we talk about, that economy will have success in proportion to the love built into the system, allowing it to evolve based on the energy of love. This love will form the creative center of the economy and potentiate innovation, thereby making the economy the most efficient. We will have to accept that those features that lead to love are dominant evolutionarily--not strength, race, wealth, intelligence, cunning, size--unless any of those affect the ability to love. The energy of Love is the energy of evolution. Motive is everything.

Art, Kindness, and E

In the arts, poetry serves as the best example of a creative act. The truth of that is common knowledge, at least to poets. 'Moral Fiction,' as defined by John Gardner is another example. According to Gardner, true art contacts and changes us by moving us closer to God, but not by moralizing. True art allows us to touch the creative source through the artist's craft. Art frames a system for perception and roots itself in love while exploring open-mindedly what it should teach. In art, we find nothing to resent since we have responsibility for our own conclusions (all the above according to Gardner).

We all have art in our lives, and with courage we can express that art. Nearly everyone I meet does not see the art in their lives, and if I mention it, they are both surprised, and always grateful. And by art, I do not mean the fine arts. I will explain much more of that later.

Anyone, artist or not, can exemplify creativity in life through the expression of kindness. E would maintain that kindness has a direct link to the energy of life (and Being) and demonstrates as much creativity as any work of art. My wife, for instance, uses kindness in her everyday life as a vehicle for helping others to understand themselves. That is her art. That most-magical kindness is helping others to help others (meta-kindness, so to speak).

A kind act is not kind without the courage to sacrifice. That means many of the things we do in the name of kindness are not really kind, but more like physical contracts. We expect something in return. And when we don't get back what we expect, we feel angry or resentful. How many times a day do you see that?

People paid to have consciences, such as social workers, and people paid to provide help, such as physicians, do not have to be kind. Even without kindness, if these people really do their jobs as professionals, they can do much good. But it is not the same as kindness, and paid professionals will get better results if they are kind, too. Unfortunately, the admissions criteria for many schools in the so-called helping professions do not have a screening tool for kindness. But people *can* learn to be kind.

Sacrifice (that within the framework of the speed of light squared) frames the key to kindness. Kindness is kindness whether the sacrifice is great or small, but without sacrifice, there is no kindness. Spirit being beyond quantity, we cannot say that any act has more kindness than another just because the sacrifice appears greater. A person can practice small kindnesses during a lifetime and lead as valuable life as someone who goes to the extreme of sacrificing his or her life in the name of kindness, regardless of the publicity.

Does that mean we can quantify kindness? If so, is sacrificing your life in the name of kindness, as in battle to save a comrade, is that a greater kindness than being kind during your life?

The energy behind kindness is in the realm of the speed

of light squared (Spirit). Again, we cannot quantify Spirit. A self-serving person (a psychopath) lives for self, and any acts lack the element of kindness. Anything done by that person is a physical contract, and she/he expects to get something out of it. Plenty of us are not psychopaths, and we do things that involve other people all the time. Someone with an Eastern philosophy will say that each of our acts contain the yin and yang of existence, which means that each thing we do has some good effects and some self-centered effects. We cannot know all the effects as they disperse.

If kindness cannot be quantified, then how can it be possible to look at the many kindnesses of life and say one thing has more or less kindness than another? The spiritual element is beyond good or evil and represents that which is called holy in us. That part of any act is Spirit. The rest of the act not motivated by Spirit, remains the stuff of yin and yang, which in my opinion, is the garbage of life and dispensable. How can that be?

Simply put, if the yin and yang formulation is correct, which I believe it is, then along with the good comes unintended bad consequences. Examine some of the good things you thought you were doing for someone only to find out later, sometimes much later, that there were also unfortunate unintended consequences. Usually these are things that we might do 'for the good' of the other person. That's yin and yang, and it is always there. We have 100% responsibility for what we do, and only *we* know what motivates us. Motivation is everything.

E is beyond good and bad, and such we call holy. *A truly kind act may appear very small, indeed, but it can have a profound effect, stretching much farther and wider in our world than we will ever know.* We just have no way to measure it.

Dying for your comrade in war may appear the kindest of all acts. And it might be, because of the ultimate immediate sacrifice. If someone sacrifices his or her life for you, it changes you for the rest of your life. That is definitively positive, unless you feel guilty that you didn't die, rather than the other.

The problem is that it occurs during war, an abnormal existence, also tainted by the motive for the war which may not be the best since it is perpetrated by leaders whose motives are suspect. We enter muddy water when we allow ourselves to be the tools of suspected motives. War is an abnormal existence, and as such, it does not translate into our normal everyday lives, because it does not (and should not) fit.

The whole existence of war is abnormal, with the biggest problem being that it is shockingly intense and random. We know that random positive reinforcement is one of the strongest motivators for humans, cite the slot machine. Random powerful negative reinforcement, as in war, is equally a negative reinforcer--mostly of fear and guilt. Every death, looking back, can be figured as someone's error in judgment, somewhere (all the way from the grunt on the ground to the politician at the top). No wonder soldiers find themselves so disjointed by guilt when returning to civilian life. Any death can expose a blame. If wartime-life ever translates well into civilian life, that's the end of it all.

During an act of true kindness, both participants (the one expressing the kindness and the one receiving the kindness) place themselves in the spiritual realm, the realm of E. I am going to call this a *creative* contract, as opposed to a physical contract--one in which you expect something in return.

In the creative contract, called kindness, both participants change and both move closer to their creative centers, or E. If the person receiving the kindness does not feel this change, then she/he is not receiving the kindness in good faith, and the person is taking advantage. The person giving the kindness then appears a fool to others. The creative act is a kind act only when accepted within the spiritual realm from which it originated. This entire acceptance does not require mental processing. Such is Grace, or something that comes without conscious acceptance, but there is awareness. Saying, Thank You, is supposed to be acknowledgement of the acceptance of kindness.

A fool is a giver who misjudges. If a giver makes a mistake (and is taken advantage of by the receiver who is looking only for self-gratification) the giver should learn from that experience. If the giver does not learn, then the taker can destroy the giver, which is the case of many spouses of alcoholics or drug addicts. Easier to join them rather than struggle with it.

A fool is one who fails in sacrifice, one who is victimized. A fool, with best intensions, hurts himself/herself, or others. A foolish act derives from mental defect, inexperience, or personality defect. With personality defect, the same mistakes continually repeat. Organized religion can help such people by providing a counterforce to habitual mistake or misdeed--much better than the prison system.

A giver learns from mistakes and over the years becomes wise. With wisdom, the kind acts move on target. Having a family, and the obligations involved, motivates us not to squander resources.

Obligations are responsibilities and part of our everyday lives, but they do not necessarily need to carry the creative contract. We have obligation to those at home, mostly as a parent or as a child. We need to live up to our obligations. Kindness, however, is a step beyond that.

Experience builds the confidence of wisdom, or it can erode confidence into suspicion or paranoia. We consciously force suspicion or cynicism away, because it spoils a kind act, and the spirituality disappears. If anyone says that she/he can see the hand of God moving in a purposeful manner in any way to bring about a result in his or her kind act, then the spirituality of the act disappears. To the physicist, this is quantum mechanics in action: If you measure, the system disappears. A kind act involves the element of mystery of the unknown, a mystery inherent in any creative act. It is the quiet knowledge that positive (or evolutionary) change derives from the flow of energy (E). That is Consciousness (much more about that later) and that energy is immediate, not depending on the speed of light. Such immediacy could directly relate to the quantum uncertainty principle, certainly with quantum entanglement. Much more on this later.

Is preaching, trying to convert people to something, a kind act? It might be kind, if sacrifice is involved. But there is a definite line between the effort to educate and the effort to convert. *When we attempt to convert, we want to confirm our own reality system.* We do this by convincing others to selflessly join our conception. Conversely, when we educate, we build awareness of aspects of the universe to those who want to learn. In that process, creativity is enhanced, and the person learns about her or himself. Know thyself, as Socrates said. That's the bottom line.

Break time?

The art contained in our everyday lives and how we lead our lives is the most important art of all. With each act of kindness, we practice art. It takes creativity and also craft-- that which we learn through experience. Our closest opportunities for creativity reside in our family lives. It is hypocrisy to lead a life of art outside the family and act like an animal at home, thereby sacrificing others in the name of art.

The artistry of family, coupled with good work through the expression of kindness, become the chance we all have to directly improve the world every moment of our lives; and that artistry can also be expressed in creation of business. Any kindness has a greater effect than any of us can imagine. A truly kind act will have more effect than nearly all of the acts of the rich or powerful, most of whom think that they themselves move the world. The powerful can put forth good acts, and unfortunately such acts are neutralized by equally bad acts of other powerful people. They fight it out--the fight of the good and bad-- and they use the citizenry for pawns. There is little spirituality there.

Kind acts reverberate throughout the entire culture and beyond--as I will explain later. A truly kind act, no matter how humble, has tremendous power. Our kind acts in everyday life and the work of true artists remain the most important ways to improve the world. In so doing, we set the foundation for the possible success of a new, evolutionarily more spiritual human species. That last sentence says it all.

E and Celebration

Let us say that we give E the benefit of the doubt as a help in understanding the world around us. If we do, where in the world would E fit within traditional religious celebration? I am putting this bit on celebration into this chapter only because celebration might seem the most difficult place into which to fit E. Added to that, celebration was also part of that initial Raincloud experience.

The religious celebrations of today mostly originate in prehistoric times with astronomical events and the change of seasons. Organized religion has usurped these times of celebration and inserted their own themes. E maintains that it can accept all religions and that a person can believe in the energy aspects of E and celebrate within the religion of choice. Allow me to test that hypothesis with the Christian celebration of Christmas.

I will first briefly explore the origins of Christmas. Anyone can find this information in a general reference source.

Before the Christian era, the time of Christmas was celebrated as the winter solstice, the time of the rebirth of the sun, the time that the day begins to lengthen, the time of the return of the source of energy. They lit bonfires at the solstice. From that we derive the tradition of the yule log. The ancient Druids thought the evergreen mistletoe contained magical powers, and the Celts believed that evergreens, as well as holly, symbolized eternal life. Christians, considerably later, believed the berries of holly were once white, but the blood in Christ's crown turned them red. Hence, we have the evergreen wreath and red berries.

Dutch settlers carried to Europe the tradition of gift-giving as there lived, in Asia Minor, a Christian bishop named Nicholas who distributed gifts. Santa Claus, the mythical jolly St. Nicholas, did not enter the public mind in America until Washington Irving popularized him in a story in 1822. In that same year, Clement Moore wrote *'Twas the Night before Christmas*. The Puritans didn't even celebrate Christmas, and it took until 1856 for Christmas to become a legal holiday in the U.S. All that is history.

Santa Claus has flowered into an entertaining myth that does no apparent harm. During Christmas, parents can lie to their children and feel good about it. Children will believe the myth because of their limited experience. To children, Santa Claus seems as plausible as anything else in this seemingly magical world. The myth of Santa Claus serves as a practice-run in myth acceptance, played as a game. In turn, merchants confiscate the myth to represent a time of buying, which is to them a time of great material reward. The myth fuels capitalism. Gotta make a living after all.

Christianity has sustained the idea of gift giving at Christmas, that God so love the world that He gave his only Son, Jesus. Christmas celebrates Christ's birth. During Christmas, according to religion, God placed His energy into a human body. Yet, we are all embodied with the same Energy or Spirit. Therefore, we are all God's gifts, according to E. The gift is life.

Jesus was a prophet, a philosopher, activist, and the figurehead of a later religious movement. He was designated the Child of God by the later Church. What does that mean?

Having a line directly with God means cutting through all the layers of existence and contacting pure essence on a universal scale of Spirit for a sustained period of time. To exist at that level and still live on our earth would seem to a scientist an improbability, because with that contact we would find ourselves in the realm of the speed of light squared, a dimension truly out of this world as we know it, and on a continuous basis.

If anyone could have been at that level of existence for a sustained time, then Son of God would fit. But we really don't know what went on. We have to take someone's word for it all, and that word comes from those who were creating an organized church hundreds of years after the fact. True it may be. With no evidence, however, faith must suffice. Nothing the matter with that.

Since we all partake of the fire of Being, E would say that we all are children of God or the essence of energy. Buddhism designated as Buddha other historical figures after the original Buddha. In Judaism, the pronouncements of the prophets of the Old Testament are believed to be one with God. Islam honors Mohammed as God's prophet. Each culture covets it Buddhas, Christs, or Prophets. Each religious tradition rightly honors its original prophet as a Child of God. E accepts all that. Does that make Einstein the prophet of E?

Once again what E does not accept is exclusivity. When any group or any person begins to talk about something possessed that no one else possesses, then E begins to shake its head. If, for instance, we wished to consider Jesus as the Christ, we can focus on the energy component of the Christ: the embodiment of the energy of love on earth. This is an energy available to all—*not an exclusive commodity*. E looks on everyone the same, all in the same boat, and all with the same potential energies. We hook into universal energies in the same manner. E merely describes energies.

When we celebrate Christmas, we enjoy the present, and we look to things for which we give thanks. We can appreciate not having been in an accident, or suffered disease, calamity, hunger, or the death of a loved one. Even if we have experienced those hard times, we still find much to be thankful for. We can look to the future in a positive manner and look inward with gratitude for the gift of life. Christmas is a time for good company, friends, and the blessings of children

Astronomical events continue as appropriate times to celebrate the eternal features in humans. These are times to honor the holiness in every human and to honor the great holy persons of our cultural tradition. We also honor the prophets of any faith or culture. In religious terms, all prophets are children of God as are we in a less developed extent. E allows this acceptance.

A person can sprinkle the Christmas holiday with E and still celebrate it as a devout Christian. The traditions of all religions differ, but they can fit with E, even the candles, lights, gifts, evergreens, processions, and ornaments, whatever. Most of us would not derive much meaning or fun from celebrating the solstice, because we have no immediate tradition of celebrating solstices. We *do* look back on our Christmases with a smile, and we find ourselves going along with Christmas (despite religious affiliation), even if the thought of drudging around a department store, or trolling the internet, or listening to another Christmas commercial sends us into a frenzy.

ELEMENTS OF HEALING

I cannot prove to you that healing energies exist, but for a start, I am going to assume that they do. Assuming so, then would they be part of this world, the world that we sense or measure with our instruments? Has anyone proof of healing powers? Have you ever seen, heard, smelled, touched, or tasted healing powers? I will assume that you have not. Other than through the senses, the only other power we have is mediated through our thought. What kind of thought might that be?

We have what is called concrete thinking, roused by all our worldly concerns, what we believe we need to do in life. We need to go to work, walk down the hall, pass the paper clips, talk to other people. Desire is connected to those activities. In the previous chapter I described how the emotion of desire intimately connects to nearly all our thoughts in some big or little way.

Logic is another type of thought, perhaps more complex than concrete thinking. Mathematics is logical. Our casual thinking may be tainted with desire, but we might resort to logic to convince someone of something to prove a point. Certainly, we try that with our spouses. Do they believe our (good) logic? Dream on. As Mark Twain once said, "You can't reason someone out of something they weren't reasoned into."

Beyond logic, we might think about abstract concepts such as philosophy. We might ponder the nature of being, or truth. This is abstract thinking. I will further define it later on.

Finally, we come to the most elusive and subtle form of thought, called intuition. Intuition is distinct from other forms of thought because intuition just happens, as if the computer had reached a conclusion while the operator was eating lunch. Yet intuition remains the basis of many of the giant steps in science, and it has significant importance in personal growth.

Every day we use concrete thinking, also logical thought, and occasionally abstract thought. If any of those types of thinking were associated with healing powers, it seems that we would know it. We don't. In addition, we know that healing powers are purposive; otherwise, healing could not exist. That rules out intuition. I have just ruled out all the types of thought so far discussed as a media for healing powers. We need to look elsewhere.

If healing is associated with an even higher, or more subtle form of thought, then only one other possibility exists. The healer must act as the vehicle of the higher form of thought or power, and perhaps the healer's own thought directs or modifies this power in some way. Indeed, healing must exist with some sort of purposiveness, and healers confirm that they serve as vehicles for powers that originate outside themselves. For now, we will go with that.

Consciousness

Powers of thought that exist beyond ordinary thought are called higher levels of consciousness. What *is* consciousness? Webster's defines consciousness as, "awareness of one's own existence, thoughts, and surroundings." I like Webster, and I will use that as a beginning, and I will examine the important words in that definition. Hang on, this is a very different approach.

First, **awareness**, what does that mean?

To be aware means that the brain is processing some type of information through the senses. We know the five senses. I am going to add a sixth sense, that of the mind, since the mind (or something, whatever you want to call it) must function as the vehicle for any higher powers of consciousness. So, awareness means that we are tracking something either through the five senses and/or the mind.

What about, **one's**? What does that mean? I suppose that I could just say that one is one, and I am one. What is so complicated about that?

True, one is an individual. And individuals are many. If there are many, does that mean there need be, in the beginning, only one? We only have two choices. Either many come from many, or one produces many. If we suppose an origin of anything, then an ultimate One (or unity or singularity) is a logical certainty. Then to accept 'one's' means also that we must accept One, a beginning One. What I am saying seems kind of off base, except when you think about the word, one. To accept one, for instance, myself as an individual (me), that one can only exist because of ONE (unity), ultimately somehow contrived. Are you feeling a little contrived? Maybe, maybe not.

How about **existence**? Existence means to have life. Life is One, the original Oneness, expressed in matter. We can call it One Life. If the Big Bang is your Oneness, then all life traces back to that.

Everything, therefore, has life. Some life has more awareness than others. Life pervades all matter, just as energy pervades all. One, as the ultimate unitary, is the source of Spirit, Spirit being the realm of the speed of light squared, which when related to matter, gives us all the energy there is-- Einstein's equation again. That's a lot to think about. I can't explain it any better.

Perhaps this is another of those times to pause.

Only humans have a sense of existence, the sense of 'I am-ness,' our sense of self such that I can refer to myself as 'I,' a unique person that I am. A step below that 'I am-ness' is instinct, seen in animals. Below that we find the plants. The plants have enough sense to watch out for themselves pretty well. Last on the list are the minerals, and awareness in minerals is much more minimal, but they at least have molecular and atomic elasticity. Minerals do change in reaction to their environment--oxidize, crumble, resist rain, wind--not much, but more than some politicians I have known. I should put a smiley face here, but I won't.

Self-consciousness comes from Life, and the consciousness we see in all the other kingdoms of nature has a stepped down version of self-consciousness, but we only see a sense of 'I' in humans—maybe other primates, but I know little of that. We humans are special, and that higher level of consciousness gives us more freedom to move, react, change our environment. All this freedom we must repay through our unique burden of responsibility. With freedom comes responsibility. We all know that. We all choose to ignore that responsibility from time to time.

Webster's definition of consciousness "awareness of one's own existence, thoughts, surroundings" refers to the ability, through thought, to sense with the five senses the surrounding world, and it includes the capacity to sense through the mind that each individual forms part of One Life. Simply put, *we all form portions of one large functioning system, and the level of consciousness relates to the extent to which we realize that.* All is analogous.

The concept of 'I am-ness' exists only when we consider Spirit and Matter separate, like comparing Matter and the speed of light squared in Einstein's equation. When we say, "We are," matter and Spirit become more unitary. With "All is" we are back to unitary, or ONE. Our one, ourselves, depends directly on the universal ONE, the realm of the speed of light squared.

When we refer to levels of consciousness, this refers to how far alienated we are from the knowledge that each of us forms part of One, such being a high level of consciousness. A low level of consciousness results in a life of desire that regards only matter as real and acquisition and control over matter the purpose of life. As we ascend levels of consciousness, matter becomes finer, and Spirit more apparent.

We cannot picture in words those levels of consciousness we have not yet experienced, because our words of description refer only to what we know and have seen in our world. Existence in which we perceive matter differently could only be described by someone who had experienced it, to one who had also experienced it and been prepared for it. If we were leaves on a tree, what possible conception could we have of the ground below, even though without soil we would not exist? For now, we use analogy and symbols as the keys to the universe. Poetry comes to mind.

Time for a pause?

The Mechanics of Healing

I am defining healing as power in the direction of the life force (however you define that), originating in One, but expressed as energy (E) and mediated through thought. How could we use that power?

The realization that all is One concentrates in our minds the power that exists in higher levels of consciousness. Such realization is called compassion. If we understand that we are One, then the troubles of another bear upon us, also. We practice this in the everyday contacts we have with people. As they say, charity begins at home with the needs of your own family, friends, people you know. There are many needs throughout the world, but we cannot ruin our own lives giving away resources that might have need right here at home. Charity begins at home. Not that we cannot be charitable for other world needs, we can. But now is not the time to feel guilty, because you are not giving enough to some organization or church. First look around you. Look with open eyes and ears. And likely it is NOT money that is needed, rather, compassion, listening, time spent, a simple kindness done. Kindness also translates well into our work-life.

You can criticize what I just said by saying, Sure, you are a physician talking about being more non-material. You are near the top of the pile. How about someone who doesn't have your blessing of intelligence and education, or is suffering from the throes of war? Yes, you are right. But all I am saying is that even in the most difficult of situations, we all are surrounded by opportunities for compassion.

Concern for others demands a lack of preoccupation for the material aspects of life, boiling everything down to the bottom line. What I am saying is that the mind churns and turns about uncountable things, and they mostly relate to self. But to prepare the soil for healing, we must rein in the mind, at least temporarily.

If we compare the power of One to pure light and with the healer as a vehicle focusing that light, then the selfish concerns present in the mind of the healer will serve as obstacles, thereby refracting that light and disallowing proper focus on the intended target. An imperfect vehicle, such as someone who is trying to heal only for his or her own benefit, is heading for trouble. If energies are accessed, they won't go anywhere; and that energy, instead of being relayed, can cause damage in the one trying to do the healing. The healer will get sick, or will meet misfortune. These energies are nothing to fool around with. It is like shining a light through clear air and then throwing dust in front of the light. The light becomes entrapped by the dust. The light energizes the dust, and much of the light goes no further. Once again, motive is everything.

Trying to raise consciousness with drugs is risky business. You cannot heal under the influence of drugs, alcohol, and some medications. Certain psychoactive drugs can punch holes into higher consciousness, and those holes can gush forth power too hot to handle for a personality yet unprepared, causing damage to the healer.

If healing exists, thought directs healing powers. The healing energies themselves must travel through some sort of media, just as sound can't pass through a vacuum and must pass through a material such as air in order to propagate. In the case of healing energies, this media is called substance, and up until the theory of dark matter (and dark energy) in the universe, we had no candidates for this media. But even dark matter is coming under scrutiny with the thinking that maybe Einstein's equations concerning gravity need a rethinking. Pretty close to metaphysics now.

Regardless, healing does not require physical contact. Healing does not require knowledge or even consent of the one to be healed. But remember, only you can prove the workings of any of this through experimentation on your own.

I started this chapter by assuming that healing energies exist, and now I am talking as if they DO exist and how they work. I say that the only proof of any of this is from your own experience. If someone comes along and says that she/he can prove any of this to you, then they are a fraud. Healing is a quiet, personal power.

The transmission of healing powers depends on power of the speed of light squared, through an unknown media. That makes healing nearly immediate (beyond the mere speed of light). The square of the speed of light is on the order of 35 billion miles per second. For physicists, this speed of light squared might explain the speed of connection by way of spin between separated quantum particles when one of them is measured, though I realize that the concept of locality might be immediate. I will try to explain that later. Lots of unknowns here. And I am not a physicist. Let's go on.

To heal means that we are trying to reverse the damaging EFFECT of something acting upon the person, such as the effect of disease, accident, or mental torment. All effects have causes. For instance, smoking is a cause that brings about an effect called cancer. Healing modifies the damaging effects of causes.

To heal means to reverse a damaging EFFECT on someone's body or mind. Again, all effects have causes. Healing reverses the damaging effects of CAUSES. Healing has nothing to do with the original causes of the problem or the disease. That is in the provenance of the person. In the case of smoking, the person chose to smoke.

Three forces contribute to any result. First is the force of damage. For example, disease DAMAGES by overcoming the PROTECTIVE immune forces of the body. While the body's protecting forces attempt to counterbalance the effects of disease, the REPARATIVE forces of the body come into action.

Those forces: destruction (damage) maintenance (protecting), and creation (repair) have universal application. The Hindus call these forces the three aspects of their Trinity, though one finds these same energies in the Christian Trinity. Just different names.

Every cause has an effect. Healing changes the course of history for a person in some manner to absolve some or all of the effects. If the healer had perfect compassion, then she/he would serve as an ideal vehicle for the healing energies. The healing energies would contain enough power to overcome strong effects from strong causes. The purer the compassion, the more power involved. I suppose a person could call what happens, with healing, "miracles," but I see the process more in terms of physics: energy propagated through thought by way of an unknown substantial media. Healing adds energy to reverse the disease equation away from further destruction. See Figure (2) Disease Equation.

Looking at the disease equation, we can see that a Cause, such as disease, acts upon a person's healthy body. If the disease is successful, the EFFECT is that of a diseased body. Working in the opposite direction to counteract the disease process is the body's repair mechanisms and its immunity, plus medical intervention, plus healing. The power of healing is in the same direction as repair, immunity, and medical interventions. Healing works by adding energy to reverse the disease equation to neutralize the disease process and to begin to move the body back toward health. If the cause of the disease is powerful enough to persist, it may take more than one healing intervention. Sometimes death is a proper result. Ultimately, death always results.

Disease Process

CAUSE⟶Healthy Body ⇌ EFFECT (Diseased Body)

Body's Repair and Immunity
plus
Medicine and Surgery
plus
Healing

Figure (2) Disease Equation

If anything can be done about the cause of the acquired disease, only the person to be healed can do it. Action would be needed. That action must proceed from motivation based on the direction of the life force, One. A positive attitude would be a first step. Any caregiver has the duty to help a patient overcome negative attitudes.

Pause time?

Methods of Healing

To initiate healing, the healer must raise his or her own state of consciousness. The healer does this by withdrawing from worldly concerns to bring self-oriented mental activity to a halt. The healer must then direct the will of pure compassion toward the one to be healed. People can see light within themselves, usually inside or slightly above the head which, with will, is projected to the one to be healed. Some people use sound as a facilitation of the media to project the energies —the Om, which to many, is the sound of energy. These strategies can be done at any distance from the person to be healed. At the very least, a compassionate listening can initiate the healing process, even an understanding touch. Technique is secondary to motive. I will be writing much more about compassion later in the book.

Healing, like any skill, needs practice, with each healer individually finding the best possible way. Some people have inborn facility to carry out healing. Others can learn it just as well. People find controlling the mind the most difficult. The science of Yoga teaches how to concentrate the mind, whether it be the yoga of love, thought, or will. The word Yoga means union, union with One, a goal shared with the esoteric branches of all religions.

Study of basic principles, everyday practice of applying those principles to life, and exploration through meditation or quiet thought is recommended. Other ways exist, but none so old as the science of union, called yoga. The healing aspect of prayer is another variation that works well. All forms of healing can work better in a well-motivated group effort. Motivation is everything, and with that thought one can see why healing energies directed by you to members of your own immediate family have less effect, because what affects them, also affects you. Thoughts about self, especially those tied to powerful emotion, will pollute compassion.

Each of us has the power to heal. We are never sure what direction healing will take. As I said before, death can be a proper result. I see this frequently in my own practice of geriatrics. Regardless of the direction that healing takes, we can help a person find that direction more efficiently and with less pain and suffering. The more we learn of healing, the more effective our efforts, but no effort so powerful as the purest of compassion. The art of healing complements the science of medicine. Of all people, we physicians have the most to learn about healing. Once again, motivation is everything—I cannot emphasize that enough.

In my own practice of geriatrics, the first step in dealing with a new patient is to connect with them (and that can be done very quickly), and allow them the confidence that someone sees them and their problems and has their best interest at heart. What a relief that is.

A note of warning. People from whom you can derive help concerning healing are not those who advertise their help and/or have pictures of themselves, or have recommendations coming from other people in the media. True healers are quiet, not loud, nor commercial, nor brought forth from foreign lands with hordes of followers. Healing energies, just like the so-called secrets of the universe, are around us all the time. The skill is to focus them. I only know enough to make one recommendation: find the light within yourself, and use that light to benefit others. How you interpret what I just said will be correct for you. You can figure it out. Trust yourself, but do not look upon yourself as exceptional. We are all in this together.

DEFINING MIND

I have referred to the mind in the previous two chapters, and I have to ask the question, what is the mind? Can we identify or visualize the mind as some 'It', a thing, a spot in space or in the brain? Or is the mind the same as the brain? I will try to answer all of those questions within the reference of energy. Let us first assume that mind does exist, especially since in English we have a word for it.

We all accept the idea that we are individuals with our own bodies, brains, and thoughts. When each of us uses the word 'I', we seem to know what that 'I' is. The 'I' is an important concept when trying to identify the mind.

I might say, "I can drive a car." That means I have a body that can perform the job if my brain sends the correct signals to my muscles. Then to drive, I only need to will it. Simple.

The 'I' must refer to whatever we think we are as persons, certainly our bodies (including our brains), but also all those attributes that we each have that make us different from everyone else--the fingerprints of a unique personality.

Our personalities have left tracks resulting from things we have done during our lives, and these tracks define us as people in the eyes of others. When others look at us, they see how we posture our bodies; they hear what comes from our mouths; they see what we do. They make judgments about us when they say, "You," and they are referring to us as persons, at least to include our personalities.

When they say "You," are they defining the same thing when each of us refers to ourselves as "I?" When others make incorrect assumptions about us, we react and quickly correct them. "I am not like that!" We continually define and re-define ourselves to others, unless of course we purposely try to fool others; but I have no intention of talking about that. We feel comfortable if other people accept us as we see ourselves.

Personality and the 'I'

We accept that the 'I' consists of, at least, our personalities, validated by what we have done in our lives. I will now discuss personality in its relation to the 'I' by discussing the three building blocks of personality: emotion, thought, and will.

We EMOTIONALLY respond to situations. Is our 'I' the same as our emotions? I think we would all agree that we *feel* emotions, but I do not think that any of us would say that we *are* emotions. I am not hate or love, or envy. The 'I' stands separate from emotions, but it feels or perceives them, mostly related to thought.

THOUGHT adds another component to personality. The 'I' can think. Thinking can include all sorts of processes including memory, reasoning, imagination, mulling over past events, creating scenarios for possible future events, and then perhaps fearing the future by adding a dose of emotion.

Another form of thought we call creative or intuitive thought, and *it* just happens. Artists will admit that they may have a good idea of what they want to do before they begin, but somehow the work of art derives a life of its own; and by midway, the creative process begins to lead the artist. Creativity pours forth, just as it does with intuition and insight. Artists and mathematicians tell us that when they try to force creativity, it dries up.

The third component of personality is *WILL*. I started

this chapter with an act of will (I was about to drive a car). Thinking is an act of will, too. That's true when we 'put our mind' to something such as driving a car; but when we drive alone in our cars and our minds wander from one thing to another, do we control that? Yes, because we can turn off whatever we are thinking and purposely change the subject. Supposedly, only psychotic people in our society cannot control their own minds.

When I *will* something, then I am making an important statement that an 'I' exists. When the 'I' wills, it must take responsibility for whatever action results. Maybe *I will; therefore, I am* is as far as we can go, that whatever I *will* defines me and, ultimately, defines the 'I' in me. That may bring us closer, but it still does not explain input like intuition or creativity, because these processes go beyond will.

If we could *will* intuition and creativity, then what we define as our self, or 'I', would position itself differently. Such control would represent an expansion of consciousness, consciousness including whatever the 'I' perceives and controls. Consciousness would expand if a person could draw upon intuitive thought at will. *The level of consciousness relates to how well we realize that we all are ONE*, as I explained in the previous chapter.

We are journeying toward a definition of mind by trying to characterize the 'I' by first defining the personality and its components of emotion, thought, and will; and I can now characterize the 'I'. *The 'I' is that which perceives, through thought, and then **wills** action based on the evidence of that perception*. The 'I' must take the consequences once it wills to do something. If responsibility exists, then the 'I' remains the holder of responsibility. If the perception is faulty—as with paranoia—then that which is willed is also faulty.

Time for a pause? For some of you, my suggesting a pause is an insult, maybe? I don't intend it in that way.

Since the 'I' exists, then of what does it exist? Is it substantial (consisting of some kind of matter) or just pure energy, whatever energy is? We know that we can only define energy in relation to something substantial. Energy-*in-itself* extends beyond our ability to define unless we introduce something substantial with which to measure it. For instance, if I am standing in outer space and looking away from the sun, I could determine that space contained the sun's light only if I stuck out my hand and saw or felt the warmth of the light on my hand.

If the 'I' is 'something', then it is either energy or matter (substance). I suppose it could relate in some way to dark energy or dark matter, perhaps conceived within the ancient conception of ether. But all of that is completely unknown.

I will now explore where this 'I' might reside. To do that, I must refer to a subject that might seem totally unrelated: the concept of afterlife.

Afterlife and the 'I'

The 'I' may be a point in time and place within the brain if there is nothing after death, nothing transcending physical life as we know it. On the other hand, if there *is* life of some form after death, then the 'I' must reside beyond our parameters of time and space; and the 'I' must have the ability to transfer into that after death realm, however we might want to define it.

Let us first examine the 'I' *in the case that life after death exists*. In this case the 'I' must also exist beyond (our) time and (our) space, perhaps in another dimension of time (I will discuss time dimensions in later chapters). Still, the 'I' defines itself by *will*. The 'I' is the thinker and willer, at whatever level of consciousness. It uses thought processes to effect action. If there is a mind, then it must have an 'I.'

If we assume an afterlife, then the 'I' must exist as some

sort of eternal spark of energy or substance that makes up our inner selves. The 'I' uses the mind as a tool. *The mind consists of the elements of consciousness (love, thought, and will) as used by an 'I' as that 'I' sees fit.*

Eastern thought delineates the *elements of consciousness* as love, thought, and will. These three words merely divide the energy available to an 'I' into three categories, making up the elements of consciousness. Contrast that with the elements of personality: emotion, thought, and will. The difference between the two is that in consciousness, love (rather than emotion) is the name of one of the energies. What our personalities make of the energy of love can take many forms. Hence it is called *emotion* as an element of personality.

Any emotion has either an element of love in it, or a distortion of love. But the energy involved, is love; and love remains one of the three building blocks to consciousness (remembering that consciousness is how completely we relate to the realization that all is ONE). The energy of all emotion is love or a distortion of it. It is the personality that does the distorting.

Elements of personality: emotion, thought, and will.

Elements of consciousness: love, thought, and will.

With an afterlife, given an energy source (which we could call, God, I suppose), then ourselves (our 'I's) individually refract energy into the elements of consciousness—love, will, and thought. Put otherwise, we all have a spark of energy called an 'I' that acts as a focal point for those powers and uses (or refracts) the powers into the elements of consciousness, a packet called the mind. The mind in turn uses the brain to effect the body to produce the result. **Mind, therefore, would be the individual collection of the elements of consciousness.** Such is my working definition of mind. The 'I' exists as our own variation of the elements of consciousness put to work by willing through perception.

The personality is the individual outward expression of the mind. Personality uses the expressions of consciousness, love, thought, and will, to deal with our everyday world. We use those expressions with varying degrees of perfection. Within the elements of consciousness there is love, and our personalities can actually express that love, but mostly it distorts it into all the other emotions that we know, including hate. But the energy involved, the energy itself, is love. That means that even evil is a distortion of the energy of love, which means that evil does NOT exist as an entity separate from the energy of love. Such negates the balderdash of the devil, though it is a handy mechanism to use.

The realm of the speed of light squared (call it spiritual, or God, or energy-in-itself, whatever), when applied to the material world, as we know it, we call that energy *love*. If, on the other hand, we had infatuation with physical things, or emotions, then the results of our actions would not accurately reflect that energy. That infatuation we call *desire*. Then we are dealing with the elements of personality—what we do or pervert with the energy of love. The better we control our actions through our 'I,' the higher the level of consciousness and the more our actions reflect Oneness.

Human beings, the only earthly creatures known to have a sense of 'I' must relate to all those kingdoms below, the animal, vegetable, and mineral, with the responsibility that goes along with 'I-ness' or basically acting as responsibly as a god would act toward all creatures or things with lesser levels of consciousness. To the animal, vegetable, and mineral kingdoms, we are gods.

Now allow me to examine the other side of the coin: **the assumption of no afterlife**. In such a case, we would have no persisting (eternal?) aspects within us, and that which wills and subsequently thinks must reside in a point in time and space, most assuredly in the brain. When the brain dies, the 'I' dies. We would define the mind by the chemical and electrical reactions within the brain. Such a viewpoint we call **materialism**. The materialist demands that we prove something to exist beyond the electrical/chemical machinations of the brain. The materialist demands that we prove that something exists that we can measure.

If the mind or 'I' exist as something not included as part of the physical brain, and the mind uses the brain as its vehicle, then to prove that such a non-material (or less material) spiritual or psychic world exists would require an instrument of measurement within that world. A physical instrument, as we know it, would not suffice. Physical instruments measure only physical phenomena.

We must rely on the only instrument immediately available to us: ourselves. Only *inner* proof is possible, but it is available to each of us. Proof takes the form of experiences. These experiences we call by such names as conversion, initiation, synthesis, or enlightenment. These occurrences, validated by what we know from life through our past experiences, reprogram our lives to new realizations.

"So, what?" says the materialist.

I reply that the system or orientation that a life then courses or proceeds upon, after one of these experiences, will fit precisely into a system already thought out and carefully written by someone else thousands of years ago. I can say this even if the materialist says, likely correctly, that any great insight, or intuition which reprograms the mind is purely an electrical reorganization phenomenon within the brain; and that is just how the brain works. Even so, the most important thing to remember is that all such intuitions or insights lead to something nonmaterial, ultimately to unity (Oneness). You can only believe that if you have had one of those experiences yourself, or examined the experiences of others (which I did by reading an entire set of encyclopedias). Some of the best writing I have ever experienced was scattered within those encyclopedias.

That which we logically conclude from what we know from our own experience, but which we have not directly experienced ourselves, we call faith—faith in the validity of what we think we know, having faith that we are heading in the proper direction. Our words cannot express such experiences until we have had the experience, and even then, there comes a time when the symbolism of words fails. The validation of these experiences resides within themselves. That is the *only* validity. Holding onto faith is unnecessary.

If the experience points to a new paradigm (personal or otherwise), then the application of that new paradigm or idea will reveal whether or not it works. We approach these experiences through the creative act, with each experience being an original personal step beyond, a chance taken, a leap of faith as Paul Tillich would say. When we fail, we feel foolish; but if we learn, at least we try not to repeat the same mistake.

So, I cannot prove the nonmaterial world to the materialist. Does that mean that the spiritual world (the realm of the speed of light squared) exists and everything I have said previously is true? Is it true that the self is an eternal spark of 'I-ness' that refracts incoming energies into highly individual forces called the elements of consciousness; and the entire collection of such elements we call the *mind*; and these forces we manifest through our personalities in the world through our vehicle called the brain? That *is* what I'm saying. Is that the way it really is? Only you can know and decide for yourself. No one can prove otherwise to you.

But no materialist can deny that energy exists, and if someone wanted to call energy (or energy-in-itself) God, then there is no possible mistake. It is only terminology.

Right?

Pause time?

Intelligence and Intellect

Mind is the individual collection of the elements of consciousness. I have, above, discussed love as the energy of emotion. Regarding the energy of Will, I have dedicated an entire later chapter. Regarding thought, I still have some things to say about intelligence.

Intelligence is the horsepower that the 'I' can focus in a coherent fashion. That horsepower consists of the sum total of the powers of consciousness utilized by the 'I,' the mind being the uniquely personal collection of those powers or elements of consciousness.

Intelligence depends on an 'I.' The 'I' can call on or release powers within its defined 'I-ness,' reaching into the unknown areas of creativity where new paradigms exist. If we have an eternal spark within each of us, then we all have the potential for accessing genius in some way.

Genius I will define as an extremely high level of consciousness, that is, highly expressing the concept of One, manifested in any or all of the elements of consciousness. To have genius in all the powers of consciousness—love, will, and thought—is truly unusual, and perhaps a few people in history have been so blessed. Religions posit their originators (prophets or messiahs) as having that power. More commonly we see genius in thought, someone like an Einstein; or genius in love, someone like a true saint; or genius in will, someone like a fully developed yogi.

As opposed to intelligence, intellect refers to collecting data and tabulating and associating that data to come to conclusions utilizing logic. Intellect refers to the scientific method in all parts except the original idea that starts the process.

The computer expands our powers of intellect and can

serve as a tool for applied intelligence. The computer only has the ability to call upon high speed intellect, a level of mentality without an 'I.' Such rules out the possibility of intelligence for the computer and rules out all the great bonuses of intelligence such as intuition, insight, and creativity. Then again, the computer fortunately cannot have the blindness caused by avarice, competition, hate, and attachment (except to the wall socket).

The computer uses the two choices, on or off, in millions of potential combinations in order to solve a problem or describe a situation. A computer cannot arrive at a third choice that we call synthesis. Within Western thought, Hegel well describes synthesis as a new and more inclusive direction resulting from two processes interacting, a mechanism that he calls dialectic: thesis and antithesis resulting in synthesis. Synthesis adds up to more than the sum of its two predecessors—thesis and antithesis. Only an 'I' can grow. Only a human individual has the freedom to make the choices necessary for synthetic thought.

The computer cannot add up to more than the sum of its parts, have individuality like a human, or possess the elements of consciousness (love, thought, and will) as does a mind. To have the powers of abstraction, intuition, or creativity, any device must have individuality, which means having an 'I.'

What is the origin of the 'I' that represents the individuality of the mind? The answer to that question for the vast majority will remain a matter of inference. Call it faith. Call it the logical extension of what we know, but always fitting properly into a very old system. I will talk more of the origin of the 'I' in the chapter on Will, but for now, I might recall what Will Durant said, ". . . for in philosophy, all truth is old, and only error is original."

On to Ownership.

OWNERSHIP

At dinner with friends I made the statement that I was uncertain about the concept of ownership. A friend sitting to my right immediately replied that ownership was easy: If you had the money, you paid for the item, and then you owned it. Simple.

Isn't there more to ownership than that? Let us examine ownership in terms of energy. More than anything it is an examination of cause and effect.

Property (possessions), of whatever kind, supplies a need and serves a function, satisfies desire and thereby brings pleasure or pain. How we deal with property in many ways defines us as persons. Property has value if we need or want it. If we feel indifferent to it, then it has little value to us. We value property according to our desire. Allow me to use an example: something useful like a paper cup.

Property in Perspective

This paper cup originally comes from wood. Someone owned the wood, sold it, and someone made the wood into paper. A variety of people through their labors added value to this product to become a marketable cup. I paid for their labors and the raw materials in the purchase price. My own labor paid for their labor and materials through an exchange of relative value, called money.

Ultimately in all products we deal with nature--a tree, soil, and water--in the case of the paper cup. To bring a product from nature to our hand required the application of energy. Energy also is a natural product. We must remember to realize the energy of creation—the mental energy that the inventor and designer needed to conceive the cup in the first place. We also pay for the energy to manufacture, distribute, and sell the item. All the energies involved therein ultimately came from the sun, including the actuality of our existence.

We have a cup made of organic material—from the EARTH.

We have creative energy applied in the form of THOUGHT.

We have energy utilized from the SUN as all our energy depends on an existing sun.

We also have value for the cup's use which is controlled by DESIRE. Without desire, no one would have ever made or sold the cup.

Something rings a bell here. Earth, Thought, Sun, and Desire? Sounds like a great vacation. Actually, more than that. Here we see all the elements for earthly existence. The ancients called it with slightly different names: earth, air, fire, and water. To them, earth was earth, air was thought, the sun was fire, and desire was water (desire certainly does flow). They actually thought of these concepts in that way. Maybe they weren't so ignorant after all.

Attachment and Ownership

We seek to own. Ownership means that the item becomes attached to self. We attach value to ourselves with the item, and ownership becomes a way of evaluating what I will call personal value. He is a millionaire. He = Millionaire (now, billionaire). A person has much money, money being the currency of value; therefore, the person has personal value. Noting the number of suicides after the 1929 financial crash, it is likely that many people equate money and ownership to personal value.

Obviously, there exist many human qualities such as kindness, compassion, honesty, and harmlessness that reside beyond personal value. Those qualities remain independent of wealth. Hence, we often hear quoted the Biblical aphorism attributed to Jesus that it is more difficult for a rich man to enter into the kingdom of heaven than for a camel to pass through the eye of a needle. The Church has for centuries used that aphorism to convince poor people that they stand a considerable number of steps in front of rich people on the stairway to heaven. It reminds me of (old) country music: a singer getting rich by singing about being poor (and especially waiting in the rain for the train). The poor man will get to heaven, and the rich man can have his limo...to hell.

Wealth has something to say about self-determination of *value* (personal value), but it has nothing to say about *worth* regarding higher human qualities. Wealth can allow a person to insulate him/herself and disregard the higher human qualities, so attached is the person to the objects of wealth. The opposite is also true. The anonymity of poverty can allow a person to insulate her/himself from the higher human values, so alienated is the person from society and the possibility of acquiring a meaningful life. Both sides of the same coin.

Stewardship and Sacrifice

Stewardship means that we control the possession rather than the other way around. Ownership to most people means attachment to what is owned. My car. My property. From ownership with attachment we see people at their worst. We certainly do not see the higher human qualities such as kindness, compassion, harmlessness resulting from attachment to one's possessions. Is it possible to own something and still avoid the radioactive fallout of attachment? That is what stewardship is all about.

In order not to jeopardize higher human values by attachment, we cannot own anything beyond our ability as a steward. Ownership of property only serves as one example of the fact that we are owners of everything that we do, all our actions. Stewardship means each of us holds the responsibility to initiate a conscious realization of the results of our actions toward an acceptable result.

We cannot predict the results of all our actions, but we can try to monitor all the effects that we cause by our actions, not just with property. If I see an improper or unintended result of my actions or my possessions, then I am required to act to correct that result as best I can. Such forms a responsibility, and it might require action on my part. That action is called **sacrifice**, a form of thoughtful love.

Stewardship is responsibility for our possessions with an eye toward a greater good in relation to the use and disposal of the possession. Sacrifice is the giving of something of value or worth for a higher objective. To be a steward of anything requires possible added time, effort, or money. Only with the rules of stewardship do we avoid the negative aspects of attachment.

Time for a pause?

Examples of Stewardship in Ownership

My paper cup:

I use currency of equivalent value (money) to buy a paper cup. It becomes part of me as it is under my control. I use it to drink water. Then I must dispose of it properly, as the cup still remains part of me until it becomes the proper part of some other process *acceptable* to me. Until that time, I am the steward of the cup. After that time, I still hold some responsibility for the results, but at least I am not longer the steward of the cup. Likely I am paying someone else to be the steward (the trash man, who recycles).

My gold ring:

I use currency of equivalent value to buy a gold ring, because I want one, and I think gold rings are beautiful. Deep inside, I want to show the world that I can afford such beauty. The ring becomes part of me.

A thief wants the ring so she can sell it and buy whatever, let's say cocaine. She steals it, and now she possesses it; but she has come upon it in an illegal fashion, even though she did 'earn' it by her labor, but only if you view thievery as does a thief. She may rationalize her behavior by saying that all physicians are thieves, anyway, and society is at fault for condoning their thievery. That sort of thinking goes nowhere, at least with physicians.

I call the police, and the police do their job, and two weeks later they have recovered the ring. The thief is taken to trial. Do I have any responsibility in these actions?

I can prefer charges on this lady. Or, I suppose, I could withdraw charges and allow the police to release her. Wouldn't that slap law enforcement in the face? What should I do?

If someone steals something of value from me, several possibilities exist:

> 1. The police catch the thief, and the thief goes to jail.
> 2. I intercede on the thief's behalf, learning that she is the mother of three, and she needs drug rehabilitation.
> 3. The thief escapes detection, and I collect from my insurance, if I have any.

Once I lose the ring to the thief (or to any other process) I am minus the ring. We can look at this as the *minus value* of the ring. Before, it had a positive value. Possessing a minus value is a real possession. In fact, both the ring itself and its minus counterpart have equal value but reversed signs (much like matter and anti-matter). If I greatly coveted the ring, then I would grieve its loss.

When I lose the ring, the minus value of the ring abides within me. The minus value motivates me to a variety of possible actions, and until I find a satisfactory resolution for this loss, I could carry this minus value for the rest of my life through **regret**, resenting thieves and encouraging long jail sentences, and in the process, raise my blood pressure through the urge for **revenge**.

Something of monetary value can pass through many hands, and each person creates a history for that item. The gold of the ring has been around a long time. Maybe Cortez stole it from the Aztecs. Think of all the places that gold might have been. The gold has seen the entire gamut of human emotions. Desire weaves it way through the history of the gold in the ring. Do we modify anything if we look upon the ring as a steward might?

In stewardship, I hold responsibility. I owned the ring, and I allowed someone to steal it, knowing that many people would like to have that ring. If I am going to own something so valuable, I should have kept it in a more secure place. That's my fault. If I don't want to bother with so much security, then I should carry insurance. If I cannot get along and be happy even though I lost the ring, then I should not own rings of such value. As the billionaire said about owning a yacht, "If you have to ask the price, then you can't afford it."

What should I do? But before I delve into that, consider what I should do if I think a friend and neighbor has stolen the ring (an inside job so-to-speak). After all, the house isn't locked. Am I going to call the police? Or am I going to confront her? Or am I going to call the insurance company?

I call the insurance company. I don't want to call the police, and the insurance company agrees to pay the claim since there was no breaking and entering. I am paid the value of the ring.

Now what? I have not lost anything, maybe some sentimental value to the ring, but not really. Do I act any differently to my friend?

No. I know she is not a psychopath and never to my knowledge ever had troubles with the police. I may never know why she took the ring, but likely the answer will surface someday. She has to live with the fact that she stole the ring. That is her problem to resolve. Desperate people do desperate things. I do not know what is going on in her life to motivate her to do this, despite our years of friendship and neighborliness. So, I continue life just as before. No loss. She's the same friend. The best thing I can do for her is to leave her to deal with what she has done, and maybe she will be able to work it out. As counter-intuitive as this sounds on the surface, such is my approach to stewardship. There are other approaches.

With the lady who stole the ring to feed her cocaine habit, if you cannot bear to see her go to jail, your solution might to be to encourage the system to consider drug rehabilitation rather than punishment. That's stewardship. Other good solutions exist, but the barometer remains, as always, motivation.

My Money in a Coffee Can:

Let us say that we had some money, and we put it into a can and buried it. This is possession, but we have minimized stewardship to the extreme degree. What of that situation?

Money in a can means the money itself, not what it can buy, but merely the money, the possession of it, becomes the object of desire. That is **miserliness**. I will show the matter in another way by use of graphs.

The Desire Spectrum

We know that we can quantify desire. Sometimes we

have little desire, and other times our desire is great. How much do you want that expensive new car? Figure (3) shows the range of desire on a spectrum. The spectrum goes from no desire to great desire.

Allow me to examine qualities that show up at the extremes of desire as shown in Figure (3). First, **miserliness**. Miserliness has value because of the money in a can, but it has little worth, **worth being defined in relation to higher human qualities.**

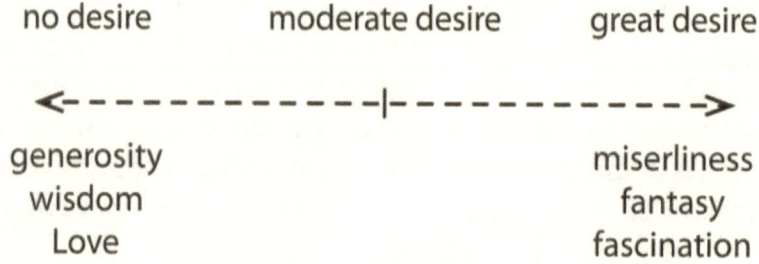

Figure (3) Desire Spectrum

Fantasy we see at the movies and on television. We also see fascination in the 'thrill at first sight' experience. We value those experiences, because they so please us; but nearly all (but not all) of them pan out to have little worth.

As we go the other way on the spectrum, desire decreases and we see the appearance of the 'higher' human qualities such as generosity and wisdom, and love. If we call someone enlightened or wise, we see a person not pulled by desire for ambition or physical objects. We remember the refrain from the song that the best things in life are free. That's an old song, but it's true enough.

Only a few people live consistently at either end of the desire spectrum. Most of what we all do resides somewhere in the moderate range. But we do experience from time to time wide swings in either direction, maybe encouraged by that marvelous salesman at the Ford dealer. Conversely, we can become spiritually charged after hearing an inspiring motivational speaker (but groaning, later, when we find out he's been arrested for beating his wife).

Further discussing the aspects of the desire spectrum, I am showing Desire compared to Value in Figure (4). Value increases as does desire.

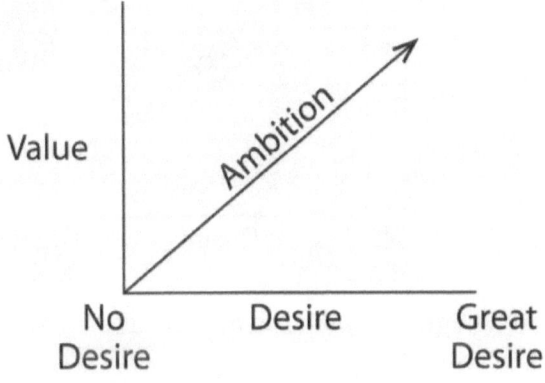

Figure (4) Value and Desire Compared

The result of both of those forces (Desire and Value) is ambition. Ambition's root comes from the Latin and means to solicit votes, as a means to power and honor (sounds like modern politics). Value pertains to money (the means to power), to the satisfaction of desire or need, and it is always a comparative amount having a number attached to it. When we ask the question, How much? We receive the answer in value. Value gauges our desire.

Next, I am comparing Worth and Desire, Figure (5) Worth increases as Desire decreases. Worth equates to what is called spirituality (the realm of the speed of light squared). An idea can have worth. An action can be worthy—as with kindness. As spirit increases, the level of consciousness also increases--level of consciousness relating to the idea that we are all One. Basically, worth concerns our motives. Stewardship has worth.

The interaction between Worth and Desire as shown in Figure (5) is an inverse relationship with the result being increasing Aspiration. Aspire comes from the French and originally means to breathe upon, but it has reference to spirit—to breathe spirit upon.

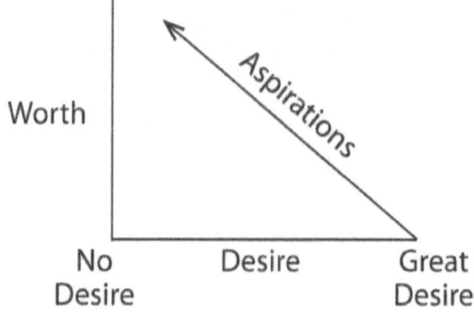

Figure (5) Worth and Desire Compared

The less we demonstrate desire, the greater we show aspiration. A good example of this is the Peace Corps. Some join it with aspiration to improve the world, but others join to add to their resume' looking toward Value in finding jobs later on—though, hopefully, they see more light somewhere through their 2 years. Some quit before that.

In our culture we tend to use the words aspiration and ambition interchangeably. Aspiration and ambition go in opposite directions, as is better seen in Figure (6) which combines. the graphs of Worth and Value compared to desire.

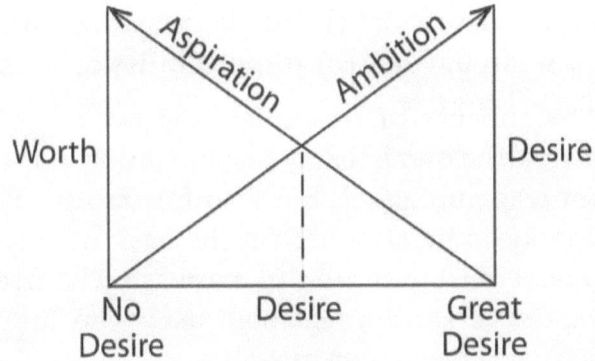

Figure (6) Worth and Value Compared to Desire

Where the two graphs intersect, we see what we might characterize as the median, or where most of us reside most of the time. Each of our thoughts and each of our resultant actions reflect a mixture of desire, worth, selfishness, and higher motives resulting in varying degrees of aspiration and ambition. We own all our actions and thoughts, just as we do property. Such involves the responsibility of stewardship, whether we realize it or not, as I will show later in this chapter.

Ownership Reviewed and More

To capsulize the concept of ownership, I am saying that ownership of property consists of possession of something through an agent of equivalent value such as money. The value of the property owned stems from the energy of our desire, desire that the ancients aptly characterized as the fluid water. With ownership, the owner possesses the item, and the item possesses the owner. The concept of ownership applies not only to property, but also to all our actions and thoughts. We own all of them.

Though higher human qualities such as compassion, integrity, and wisdom have no relation to wealth, they exist in peril because of desire. In order not to lose these qualities in the mutual possession, the owner must become the boss of the possession and take the responsibility called stewardship.

The concept of stewardship grows out of worth and implies that the owner carry responsibility not only for how the item was purchased, but also for its use. The item and the owner become attached on the basis of responsibility, a worthy relationship similar to marriage. The fate of both the item and the person are attached, just as the fate of any of our actions reflects back upon ourselves.

Obviously, I have expanded the concept of ownership much beyond merely buying something. We have considered that we also own our actions and thoughts and have responsibility for their repercussions. The concept of stewardship when applied to behavior is called **karma**: cause and effect.

Time to pause?

Beyond Ownership: Cause and Effect

Eastern thought maintains that we have responsibility for all that we do or think, and unless we act in a worthy manner, we will suffer the consequences of what we do and hopefully learn from that misery. Such a process couples with the Eastern concept of reincarnation, though it is possible to consider karma without reincarnation. Karma is physics: the energy of action and reaction. Reincarnation is religion.

According to the concept of **reincarnation**, we take what we have learned from each incarnation to the next incarnation in our effort to perfect ourselves and become the more perfect vehicles of Divine Will. In Eastern thought, heaven is that period of life out of carnation in which we assimilate whatever we have learned in the previous life; and when we complete that process, we ready ourselves for the next life. Life, itself, goes on throughout, life being an eternal expression of energy. Much of that is religion, though I will be discussing the energy aspects of life in later chapters, devoid of religion.

In Eastern thought, all human paths are equal, the only variability being where each person resides on his or her path. Some are young souls with humans acting much like animals. Others are wise and old souls. Karma works with them all, as karma expresses the law of cause and effect.

In like manner Christianity uses the concepts of heaven (reward) and hell (punishment) to motivate the believers toward correct deeds. One can accept or not accept the concepts of heaven, hell, or reincarnation, but we all must accept that when we do anything, we cause certain effects. Effects follow causes, without doubt. Such is NOT religion.

Please examine the two diagrams, the Field of Causes (Figure 7) and the Field of Effects (Figure 8). Once we have sent out the waves from our causes, we will begin to receive effects reflected back at us. Some we may like, but others will hurt. Our actions represent energy, and energy is never lost.

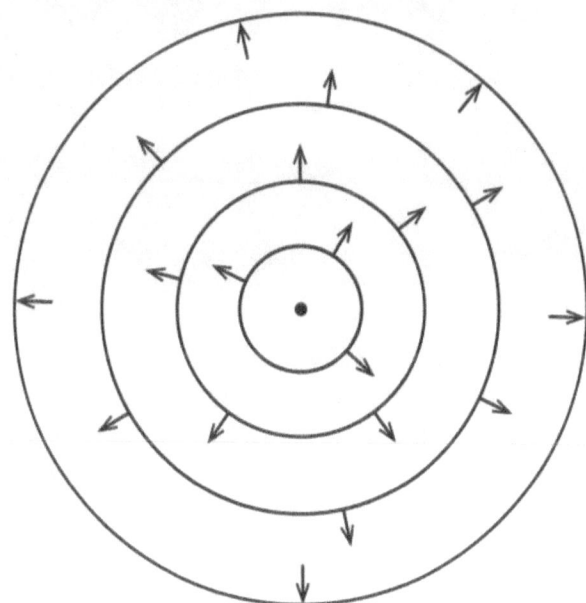

Figure (7) Field of Cause

When we focus on value, then we likely will suffer. When we focus on worth, then we have learned from past actions, and karma is kindlier. What goes around, comes around. Motive is everything regarding karma as it represents the energy we send out with our actions.

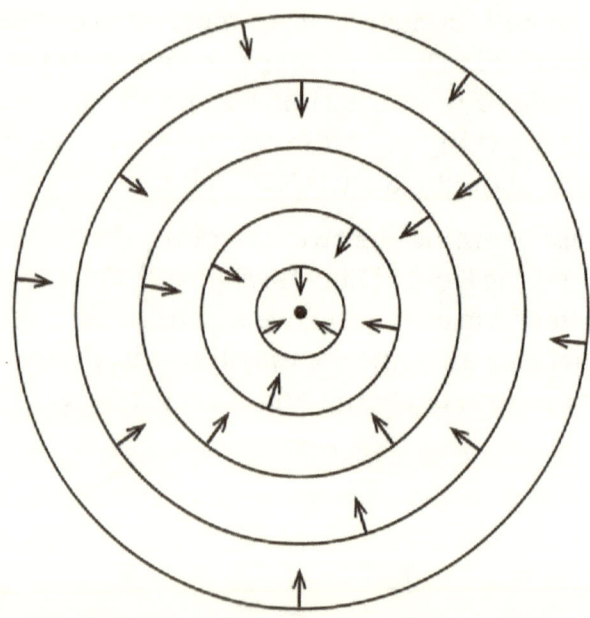

7.   Field of Effects

How long it takes the energy of our actions to come back at us depends on many factors. In Eastern thought, since there is reincarnation, karma can orbit back at us from previous lives, because we did not learn from those actions previously. Then we feel like Job, and what appears as great injustice is heaped upon us. Conversely, in people who are said to have 'old' souls, then any slip from the knife edge upon which they walk, karma actuates immediately. A selfish act then incurs immediate karma, even if minor, but always hurtful. Action/Reaction. All is energy.

The Hindus and the Buddhists maintain that when a human soul comes into carnation for the very first time, the soul comes from a highly evolved animal (maybe that cat sitting next to you is next on the list?). At the beginnings of incarnation, a human will then act much like an animal, being self-centered, cruel if necessary, instinctive. Such actions in a human will evoke harsh karma. Is that what we call evil? In energy terms, evil is the restriction of worth, just as all energy is based on the energy given to us; and that is the energy of love. Such makes evil the perversion of love.

To Do, or Not to Do

All our actions have importance to the world to some degree, but we fool ourselves that we DO anything. Nearly always we react to stimuli that come in (as from other people). We react. We may think we *do*, but we only react. Allow me to explain what appears at first to be a foggy notion.

We all react to events as they come by us, and our reactions base themselves on our history and our emotions connected to similar events in the past. How we react is a form of prejudgment or prejudice, and these forms of thought or **thoughtforms** prevent us from reacting in any new ways, ways perhaps more creative, more helpful, or ways that have more worth. I hate to say it, but we react like machines spitting out what we usually spit out when someone pushes our buttons.

If we wanted our actions to have worth, then we would have to see true benefit from our actions which only comes from reaching beyond that which keeps us running like machines, actually, beyond the parameters of what we define as *good* or *bad*—the yin or yang or existence. Rules define the good and bad. Rules do not define the worthy. Religion and philosophy do that. We fall into a pit when we legislate worthiness as is done in some countries.

Our leaders are no less machines than the rest of us. Leaders serve a broader role, but no less mechanical. They react... to polls. They rarely *do*. We have a few creative leaders that we could call statesmen. But, usually, a politician is one that only reacts—not a very useful item, but of some value.

Worth relates to stewardship and when applied to everyday life, it is called **Vision**. Vision can make the complicated more organized and meaningful. Control remains the key, that is, we must act as stewards of our actions if we want to actually *do* anything. So how would we do that?

Becoming aware

How many of us have awareness of our thoughts? The thinking of most people rolls along from one subject to another, and if someone suddenly asks any of us what we are thinking at *that* very moment, we cannot not remember it—and it only occurred a moment before! For that reason, I make the statement that we are machines, and consciousness ravels along. No wonder we react like machines.

At the end of that last sentence, what was your thought? Did your body move in response? A scratch, a rub across the nose, and uneasy shift. Can you remember the thought that set that movement into being? It's a job to remember oneself. Nearly everything that we label as being "unconscious" is actually BRIEFLY CONSCIOUS, so brief, and we nearly always make a physical movement that reflects in some way that thought. But the thought flashes by so fast, it eludes us.

Rather than the unconscious, I believe in the BRIEFLY conscious. Much more about this subject in later chapters, but I will say at this point that it is these briefly conscious events that produce the movements, such as most grimaces, facial touching, scratching, or moving the body in certain postures that we all see in ourselves and others during conversations... and when we are alone and just thinking, reacting to things we should not have said or done (I do not include abnormal movements from brain disease). So, how can we go about remembering ourselves?

We refuse to accept the fact that events *happen*. We mostly figure that we appear to have something to do with any outcome, and we rationalize when something contrary results. We lie to ourselves continually. We rarely see anything as it is. Our minds edit everything so that it coincides with our own view, and we lie to ourselves and others when we talk about our thoughts. We sleep.

Not true? Try this experiment right now.

Sit quietly, close your eyes, and try to keep you mind clear of all thought. Do it as long as you can. Experiment with it.

You likely found it impossible to keep outside impressions from coming in, noises, thoughts, sensations (have to scratch your nose, rub your face?). Second of all, if you ignored the noises, then thoughts would come streaming in, and as much as you tried to avoid it, one or more of those thoughts carried you away. The next thing you remember, you are standing at the end of a road of thought. Then you try to backtrack to where you started, and you cannot even remember that. Relax. That's the state of affairs. We all sleep.

Try it again, but this time imagine a light in your mind. You can see the light with your eyes closed, in your head, or in the front of your head, wherever—it does not matter. Focus on that light, and at the same time try to regulate your breathing to make it slow and even. Concentrate on your breathing and on the light, but think of nothing else.

That was a little easier, was it not? Perhaps you kept your mind free a little longer, but still, the most appealing thoughts came along and took you on their trips. Mind trips occupy us during periods when we are not reading or conversing. This is sleep. If we were only thinking what we willed to think, or if we could remember what occupies our minds at all times, then we would not be asleep—as impossible as that sounds.

What if we were truly aware of all our waking life, if we could observe ourselves with dispassion, independent of whatever else goes on at the same time? That *would* be awake. Would we then control ourselves? Yes, certainly. Could we then actually do something motivated from our deeper selves?

Maybe we could do all of that in time, but it is important to know that remembering one's self is an EXERCISE, no different than exercising muscles to enable us to go about our physical lives more efficiently and safely. As an exercise, it enables us to lay the foundation for self-study, and as we remember ourselves, our level of consciousness automatically rises. Like any exercise, however, it can be carried too far. The middle road is best. If you do not like exercise, then you do not like any of what I am saying. Most people hate all exercise, physical or mental. But you are different. Right? Maybe you better think about that.

Energy Control

By remembering ourselves, we begin a process, and the interesting thing about the process is that we do not even have to know the steps in the process other than see the step immediately in front of us. We find our own way, not that we find the journey easy. It takes all the will we have and more. For those who want to speed this process, and few want to, then perhaps some theory would be nice, too, but none of us should have to accept anything that we cannot realize within our own selves. We are talking about energy control.

If you experiment with remembering yourself, you will see that you can only do it for short spurts, sometimes not at all. You will note that times of increased emotion you can remember yourself better. At those times, you learn most about control. Remembering yourself does not shut off creativity or intuition. In fact, remembering one's self actually opens channels to those kinds of thinking. Successive steps appear to us *not* as large banners, but things subtle and soft, difficult to recognize many times, unless you are awake.

The directions we need to take in life appear in subtle tones, and our role is to grab them. They lead us in new directions. The drama of the Raincloud of Knowable Things only happens after we have gained many of these subtle insights through varied experience, and the Raincloud is the brain's way to bring together (or collate) related material--immediately. I have only seen this collation produce positive effects. Religious conversion is another example, and I will discuss that in much detail later. It may be possible that the brain's process of collation just as well brings together paranoid thought into a grand and irrefutable delusion, as we see in paranoid schizophrenia; but I am not sure of that.

Raising consciousness (seeing the world as One) is a learning phenomenon on a path to self-improvement. It is worth remembering that in all efforts toward self-improvement, we are always dealing with our own personality defects (Much more about this subject in later chapters). In the real world it is our personality and its defects that keep us from changing, keep us asleep. It is a mistake to concentrate on these flaws in an effort to delete them. If you want to do something about them, then try to substitute a higher motive for them, something that has worth. If we concentrate on any negative, even if that effort is to delete that characteristic, then we will only give that characteristic more strength. We must focus on the positive.

As we take more control of our thoughts, words, and actions, we would think that along the way, the hurting effects of karma would lessen. Are we not acting in a 'better' manner? Would not the resulting karma be more positive? Not necessarily so.

When we act in our dream state (unawake or unaware), we set up currents of karma (energy of cause and effect) and many of those **effects** wait in the wings to return. The more intense that energy is, the longer it takes to return. But that energy is still there. Remembering yourself speeds the process of personal improvement and raises consciousness, and as such, past karma is keyed to return, because we are going faster. It is then that the fate of Job befalls us. Previously activated karma comes home to roost (or roast)—even from past lives according to the Buddhist religion. We will suffer unless we realize and accept the justice of karma (in energy terms): cause always has effects. All education is expensive.

We eventually learn to walk a narrow path, because we know differently than we did before. We have more responsibility and also more vulnerability, but this vulnerability bases itself on the fact that past karma has been actuated, and now we are vulnerable to karma (the effects of our words and deeds) immediately. We see and feel the results of our mistakes much more immediately. Our power having been increased, the speed of reaction time within our sphere of influence also increases. We must be on our toes more than before and watch for the snake beneath the rose, as the Hindus say.

In continually working to remember yourself, over a period of time, consciousness will raise. Such a job requires massive amounts of will and courage. Is this not true of any great adventure? This way, called the Fourth Way, may have attraction to only a few, but for those few, this way will become their greatest adventure, ever. The concepts of being asleep and the need to remember oneself I have taken from Ouspensky. Google the 'Fourth Way,' and Ouspensky.

PLAN

In the previous chapter I showed how we own all our actions and thoughts, as well as our possessions in a process called stewardship. I showed how we learn from the energy derived from the law of cause and effect. We will now explore the organization of energy. If we talk of organization, then we must speak of plan. If a plan exists, then there exists a Creating Planner (big P), or at least a self-perpetuating energy that expresses itself within the laws of physics. Regardless, I can safely say that it is not all chaos. I use (big) P to denote the plan of the universe and (little) p to denote plan within our individual lives

Somebody might stop me at this point and grab me by the collar and say, "Look, Ninny, don't you know what you are seeking is already known? Look in the Bible (or the Koran). It's the Will of God, the Creative Entity that you are talking about."

"Okay, okay," I reply, "and don't wreck my shirt. Can't we, just for now, not take anyone's word for what you call God, and can we fashion an answer by ourselves, just from ourselves?"

"You're wasting your time and mine," is the firm retort.

"I get your message," I reply, taking a gulp.

I know that before we start talking about plan, we need to take a step back to what would precede plan. The bottom line is motive, and purpose merely expresses motive. Here's a simple example:

A nice young lady says that she wants to do something that is good. So, I ask her, what do you want to do?"

She thinks for a moment, then says, "Maybe help with hunger."

I suggest the food bank, and she enthusiastically replies, "Yes, hunger is a great problem, and I want to do my part to relieve the suffering from hunger."

The nice girl's motive was the will she was expressing that she wanted to do something which she considered good. Her purpose was to relieve suffering, and her plan was to work at the foodbank, which resulted in the action of working there. It goes like this:

Motive>Purpose>Plan>Action

Talking about the creation of the universe, I can frame the question in the following manner: Assuming a motive for creation, what was the grand Purpose that led to the grand Plan. I'm stumped. How could I ever understand or discover the Motive behind it all, i.e., everything that went into matter and the laws of physics? Forget Motive. Purpose will give us enough trouble, and I do have a chapter on Purpose later on, but first we must talk about plan (and Plan).

Plan (big P) must include ALL knowledge. We don't discover knowledge. That is like saying Columbus discovered the New World. The New World existed long before Columbus. Likewise, all knowledge already exists in the form of energy configurations within Plan. When conditions are right, we merely access Plan in the form of ideas and measurements, and whatever actions result from these ideas reflect newly found knowledge. Found is a more appropriate than Discover. Even better is Intuit. We intuit ideas. When we have an intuition, that must mean that we at least approach Plan in some little way.

Because we cannot examine all of the universe, nor can we examine all knowledge, we have to make the following very important assumption: so above, so below. Such is the route of this entire book, and that is the ONLY open route unless you want to take someone else's word for it all, such as from a so-called Holy Book.

I will try to answer many questions during this chapter, and I will use the analogy that everything connects; and it is all consistent, not chaotic. We will look just at that micro-universe within each of us. How each of us is put together inwardly must also reflect a higher organization, a Plan. Organization infers Plan

I begin with what anyone can do to effect any plan in life. What can anyone consciously do?

First, I can think, use thought to move my body, imagine things, use logic, deal with abstractions, have intuitions and ideas. I wrote about this in the Mind chapter.

Secondly, I can feel. I can have desire and as a result, feel lots of different emotions. I wrote about these things in the Ownership chapter.

Finally, I can will. I can use that power of will to push forward actions in my life.

Thought, Feeling (Love), and Will make up the Elements of Consciousness. I explained the elements of consciousness in the Mind chapter. We have the following flow of energy which puts all three of the elements of consciousness to work:

Plan > Elements of Consciousness > Action

Plan has to manifest through our will, thought, and love in every action we take in some very little, or large way. Any action must reflect Plan in some way, or the action could not exist. You may say that is impossible, but I ask you to reserve judgement.

When we use our elements of consciousness to effect action, does that mean we consciously control ourselves all the time? Sometimes we do, when we have good internal integration, but most of the time our actions and thoughts are unremembered by us and over-ridden by habits of reaction. Nevertheless, we hold responsibility for our behaviors, and the more consciousness we have of the components of our actions, the more those actions might reflect Plan.

Remember, Plan (big P) is cosmic Plan, and plan (little p) is Plan *represented in us* and our makeup. How each individual plan is part of the cosmic Plan is what I am going to call Soul. So, the definition of soul, for this book, is plan (small p), or the realization of Plan in each of us, and that is the energy we access when we access some aspect of Plan, manifested by what we call ideas. The collection of the potentials of all human souls in existence within Plan could be called the Universal Soul.

Again, I am speaking of energy configurations, as each soul is an individualized collection of energy. Esoteric science says that our world is surrounded by an energy envelope called the etheric body, and that body is the energy we access if we access Plan. As yet, we have no physical evidence of such a body of energy. We may never have.

To briefly summarize where we are now, I am saying that each human has potential access to a repository of knowledge that reflects Plan and which makes itself known to humans through ideas. These ideas we express in time and space through thought, will, and emotion, and ultimately in actions, actions for which we individually hold responsibility. Humans have plan, or soul, which defines a human's energy potential during that life, or incarnation. Both Plan and plan exist out of time as we know it.

Three Energies

We each have 3 energies to work with in our lives:

thought, will, and love—the elements of consciousness. We also see that the number 3 works well in religion: The Holy Trinity, and in Hinduism: Vishnu (the energy of preservation), Shiva (the energy of destruction), and Brahma (the energy of creation).

If we use analogy, the three basic energies we have to work with (the powers of consciousness) are part and parcel of the three basic energies of the universe. We will go with that, but we would do well to call the energy of emotion within the universe something else. Traditionally humans have called that energy Love, and I will go with that. At this point, it is only a name.

All emotions base themselves on love or a human perversion, or distortion of love—just the same way any idea must base itself on Plan to a greater or lesser extent, no matter how perversely. I will come back to this subject later, but hate, envy, regret, are emotions that we see all the time, but at the base of them is the energy of love, but greatly twisted through perverse personalities. Psychiatry has known this for a long time and has long considered the abnormal to be just an extension of the normal.

We have the three energies of love, thought, and will to work with. You may remember that *level of consciousness* was defined earlier as how close we come to accepting that all is ONE. So, if there are three energies to work with, we have to accept that this three come from ONE, however we may want to characterize that ONE. I will now examine how ONE gives forth Three.

In the beginning, we can assume the presence of energy, as all is energy. As I have said before, pure energy (energy-in-itself) unrelated to matter equates to Deity, if you are a Deist. Remember at the beginning of this book, the original insight was that if God is characterized as energy, then certain things play out. How it all plays out is what this book is about. If we look at the energies of love, thought, and will as universal energies and we step them down to humans, then accepted terminology changes to call them FORCES, and they are the forces which make up our personalities, personality being those same universal energies, now being expressed by us for all the world to see and judge. The personality mostly masks what we really are, which is the energy configuration of the soul (plan—little p).

How do our energies of consciousness relate to the powerful universal energies from which they derive?

Love is the ensouling energy of LIFE

Thought is associated with MATTER, since matter (our creations, or creations on the universal level) represents the result of thought, or what I am calling thought.

Will is SPIRIT, the energy within all matter (including us, since we are matter). Such will serve as the definition of Spirit within this book. Will is the energy within all matter, defined as the speed of light squared.

I realize that those three assertions seem vague at this point. If it makes sense to you now, great, but don't get tied up with it. I will return to that material in later chapters.

Let us approach this same problem of One differentiating to Three from another angle. From One we must go to more than one thing, and having many things. What are all the possible relations that can exist with things?

Things have relations to their insides or components. This concerns the construction of matter.

Things have relations to other things. This concerns the aspect of love, or relationship.

Things have relationship to that One, or the origination of creation or whatever created the thing or everything. This is the connection of Spirit.

There are no more aspects to any piece of matter than these three relationships, however you want to characterize them in your own mind. Realize the billions and billions of things, from atoms to galaxies, and they all have only three relationships, and they all require the actions of three energies to make them what they are.

Does all this mean that one moment we have Three, and the next moment we have billions? Some Christians believe that: instantaneous creation of everything all at once: the differentiation of the Trinity from One to Three to universe. No evolution. The historical record and observation of the process of change shows something quite different, but observation alone may not give us the entire answer. We have a problem, and we will return to that problem. For now, we have gone from One to Three. Now what? What comes after Three?

Let us see what three can do, in the simplest combinations. I will call the Three: A, B and C. What are the possible combinations of three?

A is one.

B is two.

C is three.

AB is four.

AC is five.

BC is six.

ABC is seven.

We have a total of seven possible combinations. If everything has to have Three, does that mean that everything must also have seven? Asked in another way, if all knowledge resides in Three, and this Three pervades all of creation, does that mean that all Seven exist in each creation, too? I will use geometry to see if it clarifies these concepts we are working with.

The triangle (Basic Trinity, Figure 9) connects A, B and C into one figure, the only way it can be done. We know that to derive all seven possible combinations that A and B must relate, as well as A and C, and B and C. A, B, and C, also need to relate perfectly together to get ABC, the seventh possibility. Since each much relate, then the first task must be to find a point of symmetry. We begin with finding the midpoints of each side of the triangle. If we extend perpendiculars in from the midpoints, we would have a new structure, as in Figure (10).

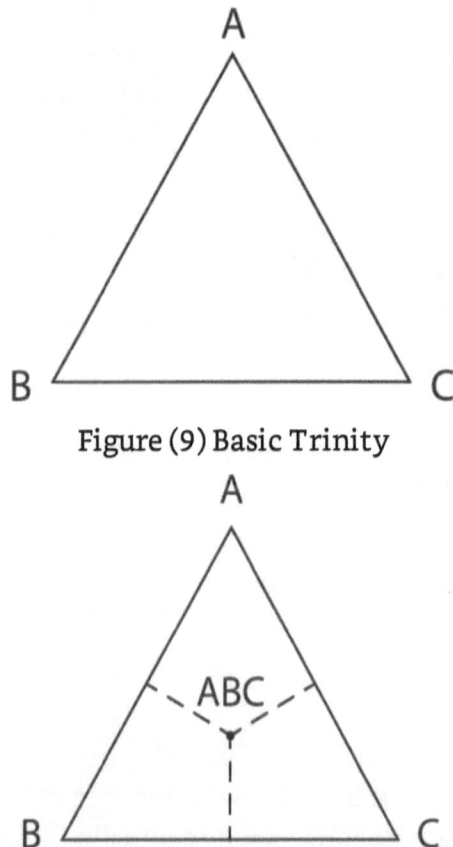

Figure (9) Basic Trinity

Figure (10) Triangle with Point of Symmetry

In Figure (10) we have defined a point in the middle of the triangle. This point, central as it is, is equally distant from A, B, and C, labeled ABC. We now have 4 points, with three more to go. We can extend the perpendiculars out from the point of symmetry, ABC, the

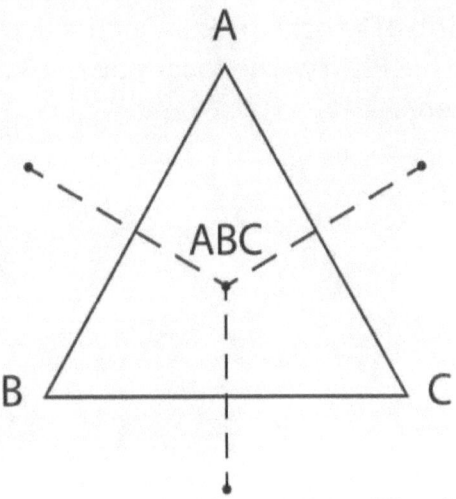

Figure (11) Triangle with External Perpendiculars

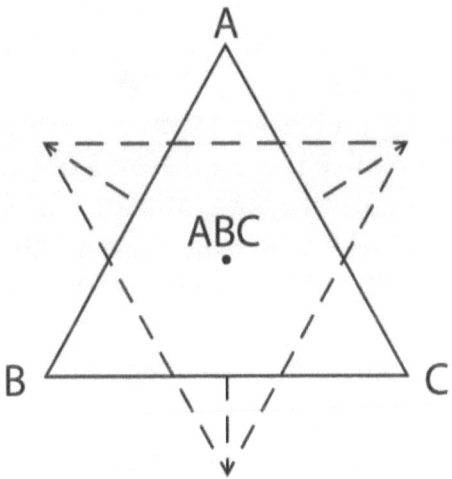

Figure (12) Six Pointed Star

same we have done in Figure (11).

We can now connect these new points as in Figure (12). We have a six-pointed star with six equal and related points and each equally relating to the point of symmetry ABC.

In Figure (13) I label all the points. One, the point ABC, has manifested into Three, which created the foundation for manifestation of a total of Six, which with the manifestation of the Creative Entity (the focal point ABC or ONE) makes a total of Seven.

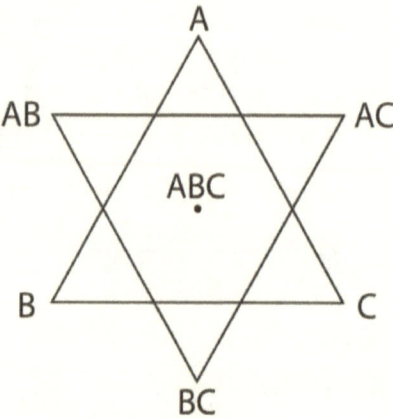

Figure (13) Seven Energies

We can do the same with light. We have three primary colors: red, blue, and green. Mixtures of blue and green make cyan. Mixtures of red and green make yellow. Mixtures of red and blue make magenta. Mixtures of the three primary colors of light make white light. Remember we are working with light, not the colors we draw with. Other mixtures of the three primary colors can produce every other color of the spectrum. A mixture of cyan, magenta, and yellow (the subtractive primaries) make black.

We have the trinity of the primary colors, A, B, and C which when mixed make white light: ABC. When mixed in duos, they make the subtractive primaries, AB, AC, BC. Altogether we have seven, as we see in Figure (14).

In light, as well as in the universal creative energies, we

can see seven. You will have to decide upon the validity of all the above by yourself, since symbols, though meant to be universal, are also a very personal thing.

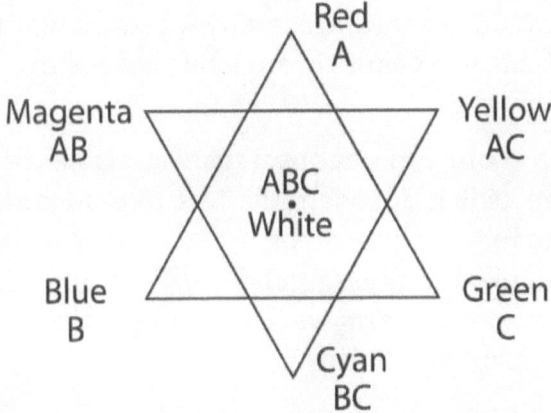

Figure (14) Seven Colors

From One, to Three, to Seven. From the Seven, thence 49, and by the same process, billions. The Seven energies make up the beginnings of the qualities of life and appearance as we all see them. Plan is manifested. The figures that I have shown merely represent the geometry of the manifestation of energies. Geometry is something physical that we can relate to. Prior to that geometry, I couldn't say, but it is comforting that the presentation of it all can be grafted. Comforting, too, that all is not chaotic. From this geometric beginning, we move to examine the process of change.

Pause time?

Evolution of Form and Evolution of Consciousness

We can say that the Primary Seven energies go into forming what we see today. What about tomorrow? Everything changes. Plan is about change. So how might the energy configuration within Plan result in change? We can see two lines of change (or evolution) throughout the world:

1. The evolution of form.

2. The evolution of consciousness.

Both of these lines of evolution link to cause and effect. In the case of form, nature does her thing with volcanos, floods, pests, diseases and thereby forcing adaptation. Change in nature can also occur through human action, like digging a hole, for instance.

The evolution of consciousness also extends throughout nature as it links with life. Life pervades all. Consciousness exists in its primordial state within atoms and becomes more recognizable as we ascend the kingdoms of nature. Once within the human realm we find that consciousness can begin to evolve through intention.

Any human must have some of each of the Seven primary qualities of energy, remembering the 6 energies and also ABC, the primary energy which is Oneness. The evolution of consciousness in each individual depends on other souls, too. The energy of love is a grouping energy, and people work together to apply intelligence and further evolve consciousness. We receive ideas from others and also de novo from inspiration and intuition. What we, individually, might see as truth may be truth for us, but only part of the answer, the other aspects of this bigger picture supplied from the actions of other people. We all need and learn from others, whether that is from reading, personal interaction, or just observing other people.

I am going to take the time to examine how the process of evolution of consciousness works within an individual and also within groups (the evolution of form I will treat in another chapter). It is a little detailed, but I will shorten it as much as possible. All in all, the evolution of consciousness is the same as the evolution of ideas, whether that be within an individual or a group.

Paths of Evolution of Individual Consciousness

First, I will show the seven possible paths of evolution of consciousness in an individual. Each of these paths actually has a name, but I think naming them adds useless detail.

1. A person has or hears an idea. The idea appeals to the person, maybe not knowing why it appeals, and a direction appears. The person begins to experiment with the idea, without really understanding the implication of the idea.

2. Understanding occurs. The person suddenly sees the idea and can describe it and how it fits into a bigger picture of life that the person had not grasped before.

3. The person begins to lead life differently as the idea has modified her/his existing conceptual framework. Now it is called an ideal. The person begins to lead life differently.

4. The person might build a conceptual bridge between old concepts and the new concept thereby demonstrating a way of change. Old concepts must be left behind.

5. The person might want to prove scientifically aspects of the new idea.

6. The idea may be applied to some form of art, including an artful lifestyle.

7. Finally, the person might want to work consciously within a group of individuals similarly interested in the ideal. I point out that this particular energy reveals the energy of ABC, or Oneness.

Paths of Evolution of Group Consciousness

Once in a group, the energies change into what we call group dynamics. Again, I will not name the energies involved.

1. An attraction between individuals motivates these particular people to come together in a group.

2. The individual must realize that coming together has an advantage which will express a result greater than the sum

of the parts. The individual must sacrifice to the good of the group.

3. The group forms a goal, which represents a synthesis of the ideals and desires of the group members.

4. The group excludes influences contrary to the goal decided. This exclusionary energy is a form of discrimination.

5. A plan of action comes about through give and take between the members, and in the process group members personally grow.

6. The group must carry out the plan to its natural limits with necessary modifications along the way.

7. With the goal approached or reached, the group realizes unforeseen extensions of its work which will require further effort of a larger scale. New, more inclusionary, groups need to be formed. As with evolution of consciousness of the individual, this last aspect of evolution of group dynamics is the energy of ABC, Oneness, always bringing more together.

The course of evolution of consciousness, be that of an individual or of a group is always towards more inclusiveness. I will say here, without showing evidence, that the evolution of form runs parallel with the evolution of consciousness. Later, I will try to clarify that statement. For now, it is worth remembering that we are seeking Plan through examination of how individuals successfully carry out their intensions and how that process, through analogy, must reflect universal energies stepped down to individual energies, the difference being only a matter of scale.

Pitfalls in the Evolution of Individual Consciousness

We approach our plan (the energy configuration that powers us—our souls), but to do so efficiently we would do well to have our tools in order and good repair. The elements of our personalities must have integration and work under conscious control.

We should not allow thought and feeling to pull in different directions. Action must reflect conscious thought. When action might be unconscious--"I didn't know what I was doing!" --then we see neurosis (though, as I said earlier, I do not believe in the unconscious, only the *briefly conscious* which is too much of an instant flash for most of us to identify). Though I will examine and define neurosis in detail later, I want to divert a few moments to show how it can serve both as a stumbling block and a tool.

We physically express neurosis through mannerisms: the insecure smile, the brief flick of the nose, the grimace, the shutting of the eyes, and numberless other movements. We look for these mannerisms when we try to gauge the truthfulness of a person with whom we might be talking. We have discomfort when a person unblinkingly looks into our eyes when we are talking, as we fear that person might have the power to look within *us* and reveal our own defects. We become guarded.

Self-doubt is the cornerstone for neurosis True enough. Though if we look upon neurosis as part of the human condition rather than evidence of some psychological disorder, then we might see something different. When we doubt ourselves, we give ourselves an additional perspective to what we are thinking or feeling which consists of what we think we *ought* to be thinking or feeling, usually related to someone else who thinks or feels that way (and we envy—or maybe hate —them for that). But with these two perspectives—what we think we are and what we think we should be, even though one of them is created through doubt, we allow ourselves a form of stereoscopic vision, and with this vision we can begin to learn about ourselves. Without that second perspective of self-doubt, even though it is called neurotic and hinders growth if adamantly adhered to, it is the very beginning of growth.

The Individual and Groups in the Evolution of Consciousness

There will come a time when self-doubt lessens greatly, and usually it is a rather rapid change. The process is called personality integration, also called becoming real, and it is just another of the processes or initiations in the evolution of consciousness in any individual. Once that occurs, what are the next steps?

We can have clues to plan, maybe Plan, with intuitions. The intuition process is rightly called synthesis. Is a Raincloud of Knowable Things an intuition, or merely a connection of ideas held in separate parts of the brain brought into consciousness as coherent ideas? Yes, it is that, as I stated in the first chapter. Yet it is more than that. The 'more than that' is synthesis, a process well defined by the philosopher Hegel.

Synthesis is a bringing together disparate views (the thesis and anti-thesis denoted by Hegel) into a more inclusive and truer compilation than what the previous views were pointing to. A higher truth. An ultimate truth? Maybe, but then that would be pointing to Plan. Mostly it points to plan.

The most absolute of human realizations of truth is to access some portion of Plan—the compilation of all the souls —all knowledge, the Universal Soul. Up to that level, truth is relative. And it may be relative even further if we are only a portion of another universe Just because truth appears relative does NOT mean that there is not a concurrent instantaneous truth, as I will begin discussing later in this chapter.

Viewing from our personalities, we could have clues to Plan when we have ideas or intuitions, but these are unintentional contacts expressing themselves through the alignment of circumstance that results in a brief flow into our waking consciousness. The vast majority of people do not have the ability to consciously access Plan; plan, yes, but Plan, no. Perhaps that is good since we humans, with few exceptions, cannot yet even control our personalities.

Somewhat aside, let me say that the knowledge of Plan, even plan, has danger in the wrong hands. We could call the knowledge of Plan hidden or veiled. The word 'occult' has a bad reputation as a word, but it best describes such knowledge. Occult merely means hidden or veiled. This knowledge becomes wisdom when applied to everyday life after a process we call understanding. Without love, however, any knowledge poses a danger, and such is the situation with which most people relate the word occult. So be it. We will leave the word occult to the so-called 'black magicians'—a category that may or may not exist—and something I know nothing about. It may all be paranoid ideation.

Continuing to examine individual expansion of consciousness, our particular plans or potentials connect to similar plans of others. I showed the seven categories of both individual consciousness and group consciousness. The basic energies are the same in both. In everyday life, we have groups: families, friends, work related groups. But there also exist soul groups. What are they?

I have defined soul as plan—that energy which makes up our existence or energy as a human, and it is our potential. There are other people or plans that match closely our energy combinations. It only makes sense. These people can be scattered all over the world. When we access plan, then we also access plans (the similar energies) of other people. That grouping is called a subjective group or soul group. That connection is beyond language, and it is *instantaneous*. We would have good fortune, certainly an unusually good feeling, if we actually encountered a person from our own subjective group, and you would know it, even if it were the briefest of encounters. You would treasure that encounter your entire life and wonder about it. Have you had such an experience? They are not uncommon.

The laws of energy that would rule knowledge acquisition via a process like intuition are the same laws that are involved with group process. We can also access plan through contemplation which is the highest level of meditation. Some people can do it through prayer, but certainly NOT the type of prayer most commonly encountered--the beseeching type--unless the motive is pure. The process eventually can put us in conscious contact with our subjective group, and with that boosted energy, Plan can be approached. I cannot speak further about that. I am not adept. But the type of thought used *is* abstract thought, and I know about that. Allow me to discuss it, briefly.

Abstraction and Dreams

At the plan level (soul level) we find abstractions of thought, thoughts which can realize things in the world. We might find individual direction when we deal with plan, which spells out our potential. Conscious abstract thought congruent with Plan is beyond nearly all of our capabilities.

We can have some inkling of the process of thought becoming things when we consider our dreams. In dreams, thoughts *are* things. Granted, much of dreaming involves the emotional aspects of existence. Within the emotional level of our lives we see a sensitivity to patterned and conditioned reaction, not thought. Mostly we see reflex action within the emotional level, more attune to conditioned response rather than true thought. We tend to react emotionally in the same old way to the same old things.

Though the energy within dreams is primarily emotional energy, there can be some thought, and it usually reflects experience. You just saw a dinosaur movie, and one of the characters was a dinosaur in the dream. The fear that we might experience in a nightmare may not so much represent fear of anything external, but the fear of an *internal* idea, an idea that kicks up an emotional response usually by showing that change is suspected. The personality resists and fears change.

We have ordinary dreams, and we also might have extraordinary dreams, which I call visions. Visions have high clarity, brilliant color and contrast. You will readily recognize such dreams. Hold them close, remember them, since there is a message there from plan. It will put light upon your path. You have been approaching something, or doing something, which has activated energy from plan. That dream can be taken as a personal message from your own innards. Burp. (Don't confuse that with gut feeling.)

Evolutionary Change and Synthetic Change

We can have indications of plan, maybe even Plan, with intuitions. In the previous section of this chapter, "The Individual and Groups in the Evolution of Consciousness" I showed how the the Raincloud of Knowable Things is an example of a rather gigantic intuition, and it adds up to more than the sum of its parts in a process called synthesis.

Repeating what I said before, synthesis is a bringing together of disparate views (the thesis and anti-thesis denoted by Hegel) into a more inclusive and truer compilation than what the previous were pointing to. A higher truth. The process of synthesis is beyond any computer, and will remain so, in my opinion.

Perhaps, when we have attained a sizeable measure of the processes of evolution of consciousness, then change occurs through a different mechanism, one that we may not clearly conceptualize. Could it be that this process is something more akin to what the people who are fighting evolution have only an inkling of, and the application of their inklings only appears wrong considering what we see as the bare facts today?

Truth is, after all, relative. Truth becomes absolute only when it directly reflects Plan. Creation within Plan may be immediate (rather, within the realm of the speed of light squared—Spirit), and evolution may only represent the tail end of the process, as Ideas are eternal, and evolution is history.

The process of synthesis is beyond relativity, beyond evolution, more like quantum immediacy. When we discover that process, we will then look back and say, "Of course, it is only logical!" But we can only see the logic in retrospect. Until then, the knowledge has a veil (that's the use of the word 'occult' I was referring to earlier), and we only see darkness. The first indication of seeing that knowledge will be light.

Thoughts are things. Within the human realm, thoughts become things when we work on them. But at other levels of existence, the realization of thought into things could be instantaneous, as Creation was immediate within that realm, but not as we see it. Creation is beyond time. The same with Synthesis. I will return to these ideas later.

So, I would not discount all those people who defy the concept of evolution. They may be wrong in not accepting evolution as a process, and they may also be right, intuitively. Such is a paradox, paradox being the combination of what we think we understand with some other element that we do not understand to bring forth a result that puzzles us. When we see paradox, we know we on the trail of something important, a discovery beyond the horizon, a discovery that will likely occur through the process of synthesis. Whatever we seek is more inclusive than we realize, and includes both sides--the thesis and the anti-thesis—with the final formulation something we could not have predicted, and perhaps much more complicated.

"More complicated? Says the mystic. "No, it's simpler."

Let us take a few moments to check out mysticism,

as mysticism is a very particular example of expanding consciousness.

Mysticism and Access to Plan

To the mystic, the matter is, indeed, simple. The mystical process is one of attaining a union with the emotional aspects surrounding the idea level and feeling that experience, (usually one of great exaltation). The mystic will not have interest in personality control. The experience of mysticism bypasses the thought process, and as such, the mystic will not build toward the idea level and create a framework on which to hang knowledge. Instead, the mystic accesses the emotional aspects of that knowledge, and has a genuine emotional experience, treasures that experience, and generally rejects anyone who wants to mess with the experience.

The mystic loves the mystical experience, and it remains more of a marriage in its dualism rather than the union of yoga. The mystic accesses high energies and internalizes those energies, but not as a vehicle. Those energies stay within, and that's dangerous. The mystic will likely suffer from those energies with illnesses and self-destruction. Esoteric schools maintain that before one can become interested in esotericism (the study of the energies of thought) one must have had previous ability, in other incarnations, as a mystic.

The talent of the artist resides in the ability to access the energies of plan, maybe occasionally Plan, and express such in the terms of their craft. Artists who are mystics will suffer as do other mystics, and there will be suicide, self-destructive behaviors, and much illness. Some artists have integrated personalities, and they can access high energy levels, and they can relate what they experience through love and craftsmanship. Artistic genius combines both the emotional impact of the mystic and access to the thought of the idea level. All of that the artist puts together in a craftsman-like manner made possible through inborn talent and endless practice. Few artists qualify for this high honor.

Destruction and Evolution of Consciousness

We must all realize that ideas have their destructive tendencies. We dread that aspect of energy. The Hindus call destructive energy Shiva (Siva). The Christians denote that energy as that of Father. Japanese movie makers call it Godzilla. We have fascination with destruction, which is the natural outgrowth of our knowing that before we can see the new, the old must be destroyed. Both Father and Shiva express more than just destruction as they involve the dynamism of all change.

Much of that change produces violence. Marshall McLuhan, the first guru of the effect of media on society, said that with the introduction to a society of the powerful media of radio, the first reaction was always violence within the society. The same can be said of the cell phone and other media today. The violence can come from either the revolutionaries or the reactionaries. Somewhat later, the rest of the populace gets dragged in. The introduction of media enhances societal change, and violence erupts. But, as Dickens observed, sometimes the best of times is the worst of times. Violence rides with us, but that does not mean that other ways, rather than violence, cannot be found to integrate change. We struggle to find those ways, but the reactionary forces wage war.

Individualism and Beyond

The personal progress of each of us depends on others. As we progress, we do so by helping others. As we advance, another moves in our place. We might call this process uplifting. More commonly we call it service. We approach plan through service. Through service, knowledge becomes wisdom through understanding. The energies involved in service resonate with the sorts of energies we find at the idea level, the level of soul, thereby setting the media for acquisition of further intuitional ideas. We could call this the adventure of service. We work out our problems in everyday life through service, but our basic idea level is the group that makes up our soul-cell. We use those energies that energize the soul-grouping to find our own plan, and for the group to find Plan. Group direction predominates at the soul level.

Group direction rubs uncomfortably against what we call individualism. Control and integration of personality is a personal thing. That *is* individualism. Once that difficult task has been accomplished, we begin to approach plan. Individualism then becomes archaic, with the exception of our personal responsibility to do our thing in carrying out group direction, which is NOT easy. In fact, it is like walking on a knife blade, and any karma is nearly immediate. Please do not be afraid of that. If you are there, you will not fear it.

We have powers as individuals, but proving that power in the physical world no longer holds importance. Is there not a large measure of proving oneself or demonstrating one's own power within the concept of individualism? It is immature, to say the least.

As the world evolves, we will see more and more humans with coordinated personalities with access to soul. When such people attain leadership, and they have behind them enough followers with the will-to-good, then it stands to reason that we would see an evolutionary movement with participatory democracy with less emphasis on individualism.

In our present system, people vote on the basis of how certain issues impact them personally. Democracy moves according to how the majority perceives its own interest, which means that the minority, and the percentage points between the two may be slim, perceives that the issue will impact them negatively. Some call this the tyranny of the majority, and because of this, we have passed laws to protect minority rights. Few would object to such laws. Otherwise, the strong will rule: fascism.

The myth that each individual in a democracy voting for the good of her/himself individually, and the conglomeration of all these votes moves toward a common good for all, is just that, a myth. One can argue that it beats the hell out of other methods, and at least there is orderly movement. Winston Churchill once mused that the Americans usually make the right decisions, but only after trying out everything else. Probably true.

Democracy reflects the means to an end, many times inefficient at best. Conversely, fascism (or any autocracy) reflects the end through any means--much more efficient and much more destructive.

We must realize that when we do something only for ourselves, or how it might impact us, that such a decision will differ in tone, and many times in direction, than the decision that takes the group into consideration. In such decisions, we are thoroughly alone.

Critics of what I have just said will maintain that group orientation was exactly what the Communists were trying to do, unsuccessfully. I accept that criticism. The utopianism of Communism, as it was initially touted, promised that once the people had evolved sufficiently, then government would melt away, and dictatorship would be discarded for communism.

The big mistake in Communism, and it was a big one (some say a deception, but I think that smacks of paranoia) was that the people cannot evolve unless the system under which they live also evolves in some manner along with them. The Russian communist system had poor mechanisms for evolution. Does any dictatorship, or theocracy, for that matter?

Any person interested in the freedom of the human spirit, felt rightly trapped under communism, Russian communism would have continued, had it evolved economically, but it could not, and we can see why, since economics is the currency of energy. For the economy to evolve, which it must to survive, it requires the freedom to evolve.

We have a *slow* revolution in our country. We always have that, as we must, to improve. Concurrent with change, we have nearly continual wars going on. We suffer the openness of an open society, because we know that through openness, we will find redemption. We wring our hands when we decide to strengthen law enforcement, but then we fear loss of personal liberties. We feel violated when foreign communities appear in places we might have treasured as a child.

Regardless the issue, we suffer, but that suffering takes place within our personalities—that's the hell of it. Whereas, at the soul level, we find no suffering, and that's the heaven in it. We find a place somewhere in between. We vacillate between consciousness in our personalities (the mask of the soul) where desire and emotion hold the reins, and at other times we rise to great ideals.

If any concept shines through as critical, we might choose *understanding*. When we understand, we take the knowledge that we have acquired through the experience of service in our lives. Through the lightning bolts of intuition, we synthesize that knowledge into a wider vision called wisdom.

The Bottom Line

I have tried to show that Plan (cosmic organization) and plan (the individual set of energies, called soul) integrate into the human condition, thereby coordinating universal energies into human lives. Those universal energies (whether you want to describe them as products of Divine intervention or the natural élan of the energies within the universe that physics describes) are the same energies that we have to work with as humans to find our way and evolve. That evolution has an immediate component beyond time, called synthesis, and also a historic component, called evolution. We work in groups, recognized or not. Individually, we make our own decisions, and in that we are totally alone. We correct our individual paths having to deal with the energies coming back at us through the laws of cause and effect, called karma.

I showed the geometry of One to Three to Seven, thence to billions. The simple geometry is a reassurance of orderliness, and a refutation of chaos. To say that I know anything about the energies in the universe which result in the orderliness of this geometry would be a lie. But I do know that it is all energy, and I have tried to show that the energies we have to work with everyday are undefinable (at least by me) stepped-down universal energies. If as above so below is valid, then I am describing something real, and this is no mysterious Holy Book.

Concisely put: *Plan is revealed through the evolution of consciousness, primed by motive, and led by the process of synthesis, resulting in Being, rather than acting.* More on that later.

LIGHT

What might Einstein's equation $E=Mc^2$ have to do with conflict resolution? E is energy, M is mass or matter, and c is the speed of light (squared). We know that the universe consists of matter, energy, and light. Energy and matter can convert one to another and exist in equilibrium. Change in one reflects change in the other.

In the equation, light is the only constant. We have never found a variation in the speed of light. Intensity, yes. Speed, no. We can look at Einstein's equation in a somewhat different manner in Fig. (15).

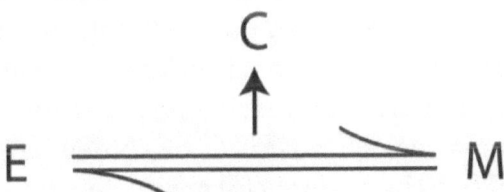

Figure (15) Metaphysical Einstein

In the above formulation, a metaphysical one rather than Einstein's which is physical, matter and energy can convert one to the other, but they stay in equilibrium. Light (C), the universal constant, involves itself in the intensity or magnitude of the reaction between E and M. We could say that light is a manifestation of the conversion of energy to matter or matter to energy. I say this is metaphysical, but only because I am showing the basic equation reversible, and the speed of light acting more like a catalyst. I do not think this has to be true—but it might be—but it will not make a difference in the points I am about to make.

Okay, what does that have to say with how we live our lives?

Thoughtforms

Every day we all live in a mishmash of energy. We have physical energy, emotional energy, and mental energy. Energies from within ourselves, from others, and from groups affect us. An example of a group energy might be something going on in our nation, a big problem the nation faces that affects us all. We all feel it. Most of the time we face shifting and interrelating energies with no mental note of it, not knowing where the feelings or the thoughts come from. Many times, the confusion and power of it all simply overcomes us. We call that stress.

Floating thoughts, directed thoughts, and emotions originating from other people affect our thoughts and emotions, and from such stimuli we form thoughts and feelings fused together in what are called *thoughtforms*. A thoughtform, or a thought attached to an emotion, becomes part of our mental repertoire, and we automatically call forth these thoughtforms given a situation similar or somewhat similar to that which created the thoughtform in the first place. In such a way, these thoughtforms become real forms, real matter to us, part of us. I realize I have made this point before, but it is important enough to keep repeating.

We tend to view reality through a jungle of our already preformed thoughtforms. If something comes along that does not fit into a preformed conception, then we might ignore the whole thing, or we might face the situation and perhaps feel anxiety and shrink from it or, hopefully, try to deal with the situation. We face any situation with preconceptions, and these preconceptions will change the way we perceive the situation. Call them prejudices or pre-judgements. Eastern philosophy calls it illusion.

A story is told that an early European explorer once arrived at a distant area in South America where the people had never in memory seen anyone outside their own. The natives had small boats and weapons, but when the white explorers came ashore and confronted the people, the natives could see the explorers, but they could not see their large ship sitting in the harbor. A boat that size could not exist. Therefore, it did not exist. I can relate to that. So many times, I try to find something, and it is just sitting there staring me in the face.

Our problem is to learn how to deal with the energies all around us. If we had clairvoyant powers and had the ability to see these energies--with our psyches as they are now-- we would go crazy. No way could we deal with all the information coming in from these mental and emotional energies since they interrelate and interact. We talk about the information age and computers. That is puny compared to this.

Ideally, we would like to have the necessary powers to discriminate, that is, to have the power to consciously perceive these energies, separate them, and then deal with them. Then we would have a conscious process as opposed to the unperceived jumble going on within us most of the time. If we want to discriminate, then we have to somehow open ourselves to these stimuli, not allowing them to overcome us. How is that possible?

Light Guides Us

The mask of our personalities is the ironclad filter tightly clasping our well-worn thoughtforms. We see what we want to see, ignore the rest. We cannot discriminate with that handicap. The first step is to consciously tag an input when it arrives and triggers an automatic response in us. We need not judge or analyze how we react, we merely need the awareness that something has arrived within our psyche.

Okay, great, then what do we do? How are we supposed to make sense of something probably very complicated coming in from outside us, not to mention everything else that is going on the same time? The answer resides in Einstein's equation. Matter and energy are in flux, but thankfully, we have light, the constant c. We can depend on light.

The amount of light involved in the interaction of energy and matter compares to the amount or the extent of the reaction going on between energy and matter. With much interchange, there would be a great deal of light. We have a good example in our sun. We could say that the amount of light (all spectra) coming from the sun directly relates to the rate and extent of the reaction between matter and energy.

I can further say that the source of light, the origin of light, has a common origin with energy and matter. Light, energy, and matter came into expression together, neither known to exist apart. Light gives us the evidence that matter and energy are interacting. Energy without relation to matter, is darkness.

The Whole and Triplicity

The energy of emotion and thought transfers from one person to another through the senses and the mind, the sixth sense. I wrote about that in the chapter on healing. In the case of healing, we act as the vehicle for that energy. But now I am talking about energy that lands within us and affects us, not being merely refracted by us to others. That energy causes changes in us, and we need to know what is going on. Otherwise, we can become sick. How can the energy of emotions and thoughts transfer from one person to another? But before that, the first question is through what media?

Any energy transmission must have relation to some sort of matter. That is where the ancient concept of ether came about. Ether is the media that was (is) thought to carry these fine vibrations or energies, just as air certainly carries the vibrations of sound.

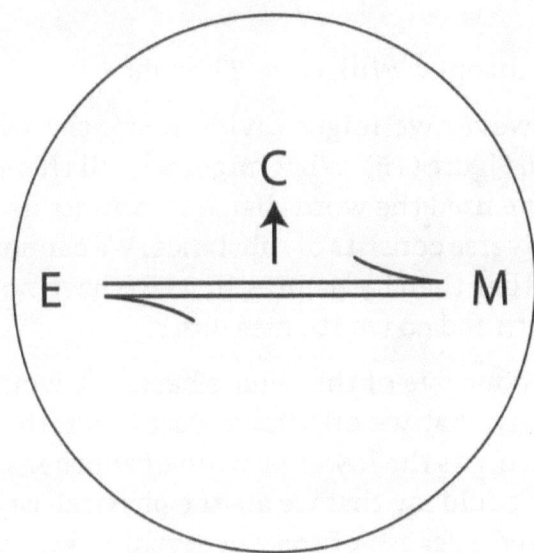

Figure (16) One System

Energy and matter do interact, perhaps as shown by Figure (15), but we are basically dealing with only one system, Figure (16), which is a truism. I say this because energy and matter interconvert, and as energy and matter relate to one another, matter changes state, as we would see, for example, when wood burns to charcoal. Nothing is lost in this change. We still find the same equivalent total of whatever makes up the matter, energy (heat), and light, though the distribution and the appearance has changed. The entire system--the *triplicity*--matter, light, and energy remain intact and contain the same total energy.

Religions and philosophies alike speak of triplicities. The following triplicities correlate directly to one another (the first listed, the second listed, the third listed). The triplicities are listed after the conceptual frameworks:

Science: Energy, light, Matter

Christian: Father, Son, Holy Spirit

Hindu: Destructive Force, Sustaining Force, Creative Force

Philosophy: Will, Love, Thought

However we might divide it up, the whole stays the same as in Figure (16). What might we call this whole? The ancients have used the word that we translate as *substance*. The entire universe consists of substance. We cannot describe this system other than by naming it, as we have nothing to compare it with and no way to measure it.

To conceive of this whole better, it would make more sense to say that we only form part of another system, perhaps existing as the lower portion of another system or hierarchy. We could say that we are the physical aspect of another system, and because of that, everything we conceive as our universe is substance. Those are interesting thoughts, but we have no proof of any of that, and besides, it merely passes the buck. How can we say anything, unless we can see it for ourselves?

What Can We See?

We can experience the factor of light. Light is a constant and it directly relates to how matter and energy interact. The relation of matter and energy changes constantly, giving us the mishmash of thought and emotional energies that we cannot easily translate or discriminate. Maya, the ancients would call it. Notice the difference between Maya and illusion, illusion being associated with our well-worn thoughtforms and Maya with matter.

We must somehow use the factor of light, the universal constant, to foundation ourselves. We know that light exists within each of us, as we know that energy and matter are interacting within us. This light must connect and link to the light of the creation of the universe. Yes, that light must still exist. The intensity of the light of creation would be equivalent to all the light that could possibly be emitted when all the matter in the universe was interacting with all of the energy in the universe, *at the beginning*.

If we believed in any way in an evolutionary direction in the universe, then that force would be evidenced by light within ourselves, actually, light within anything that exists. That light would be the remnant of the light of creation. That is called *All Light*, and within each of us that light remains as a unit of creative energy. As the universe evolves, and we evolve, we perceive further amounts of this light. Where might we find this light?

We are collections of energy. We have matter as part of us, but energies hold us together as human units. To the outside world, we consist of a physical body, and emotional self, and a mental self. We see only the physical self. The rest makes up an energy field around and within us. A cohering energy holds this collection of forces together into a living unit. When that cohering energy withdraws, we call that death. The cohering energy does not die. When it exists as part of a living human, we call that cohering energy the *Living Soul*.

When the soul energy leaves the body after death we cannot say that it is only a collection of energy since we know that energy exists along with matter. We cannot define energy unrelated to matter. There exists, therefore, subtle or fine matter along with this soul energy. That unit is called the *Soul Body*. There must be light being emitted from the interaction between matter and energy at the soul level. That light is called *the Light of the Soul*.

This light has ignition by the original creative spark coming from the light of creation (*All Light*) in each of us, and it would not have the intensity or concentration of that original spark as that light must necessarily be of the brightest intensity—Einstein's equation speaks to the square of the speed of light. But the light of the soul must have considerable brightness, and the light would come from the entire soul body. This light would evidence the energies within ourselves.

Mechanisms for Light Transmission

The energies from the soul are represented by our physical, emotional, and mental humanness. Since we have the responsibility for all expressions of our personalities, those energies from the soul only express through what we make of the soul energies, and we usually bumble that task in everyday life. No wonder we call the personality the persona, the mask of the real person.

A highly refined personality could directly represent or transmit soul energies, and a poorly coordinated personality would have elements of soul energy, since that is the real energy, but poorly expressed. Beneath the personality resides what we call character, and that is the framework. Personality is like the outer appearance of a house, and character is how the house is constructed. Soul would be the ideal plan for the house, as conceived by the architect. That plan or energy is still present to varying degrees in the final house, despite the compromises that the builder had to do to make a financial go of it, and the craft of the carpenters—some better, some worse. Soul is the light, or plan (small p), as shown in the Plan chapter.

A highly refined personality can express soul qualities through a solid framework of character. If character is solid, then the soul shines through. Those are exceptional people. Defects in character refract the energy of the soul, and that expression through the personality can sometimes be very appealing—even in a psychopath who is trying to fool people for his own benefit, or the hypocrite who is fooling himself. The color or quality of personality is supposed to be an inheritable trait. I don't know about that. Be that as it may, it *is* the personality that we see in others, and it is through character that we learn about others, mostly on the basis of the decisions they make in life and how they act.

If we use our potential for insight, we can see soul energy as the light of the soul in others. I proved that to myself in a book I wrote early in this century which showed the connection between the soul and the personality through interviews I did with townspeople, in this case women, and also in radio interviews with both sexes in an hour-long program on a local station. I will return to this subject later.

Suffice to say that light does exist and we can see it in ourselves, and we can see it in others. We only see the light in others by seeing aspects in others and identifying with those aspects in ourselves. It is sort of like a key to common energies. This process is called identification, and with identification one sees the commonality in souls, and it serves as the step just before illumination, or seeing the light, called understanding. Such a process bypasses the stumbling blocks of character and personality. Ill-starred qualities in personality or character cannot be overcome by suppression, as suppression only magnifies the strength of the unwanted quality. Substitution is the only thing that works, substitution with light. That would work in politics, too.

Discrimination, Identification and Understanding

The very first step is discriminating and identifying energies as they approach our own energy fields, from individuals or groups, or even from our own thought. To enable discrimination, we cannot have a considerable amount of emotional upset or mental torment going on inside ourselves at any moment. When we rid ourselves of such interferences, then we can discriminate receptions and see beyond a person's personality and all its complexities.

I found in my own experience how soul light can be brought to attention, not only to me, but also to the person I am talking with. Basically, it works if one focuses on whatever the person is saying in terms of a finer quality. Every person believes in their own justifications. Such needs to be accepted rather than challenged, leaving their justifications alone for the moment. You can tag or mark what the person is saying (or some aspect of it) to some soul or higher quality. If you do that, the person immediately responds positively, and they will enlarge upon that subject and be thrilled that you have identified something that they have never thought about before.

Everyone has art in their lives. That art could be something they do with their hands, but more likely something that they do so very well, maybe how they deal with certain people. When you are identifying soul qualities in other people during conversation, then you are truly dealing with art. Very few people realize this. I will return to this subject in the later chapter on Personality.

Identifying a soul quality in another person and bringing that to attention is a form of identification and then we can walk in that person's shoes, as much as that is possible, and that opens a circuit with the other person which will always be there. That open circuit is called understanding. And the emotion that accompanies that understanding is called joy. The process of expressing that understanding we call love, which means allowing soul energy as evidenced by light to flow from the soul through the mind via understanding and to express itself as love on earth through will. Thereafter our perception of the light within our own selves intensifies as does the light within the other person. This would be the route of spiritual evolution, and it bases itself on service.

The interesting thing about this process is that the light of the soul manifests better not by laying 'a trip' on someone else, in this case, our own trip, but rather allowing light to emit because of understanding. That means understanding the other person. Through that, we will understand ourselves better, which helps us in our own decisions in life. The other person would change only and exactly within his/her own energy repertoire, on his/her trip or path. In that way, we do not interfere with another's karma, as the Buddhists would say. It is not for any of us to put ripples in another's pond.

Seeing the light of understanding, is the only way to any conflict resolution. We can develop techniques to resolve conflict, but when resolution occurs, light intensification results. It takes considerable will and courage to do this. Want to try?

Creativity and Prejudice

To enter into the area of creativity, which is what we do when we access light, we must first face darkness. Darkness prevents people from taking the necessary step forward. Most people do not even approach the darkness (even to approach it is too scary), and they stay comfortably in their old thoughtforms and do not budge a bit. Faced with the blackness of a step that we have never taken before, we feel the fear of nothingness, the basic fear of death, the fear of the unknown: anxiety. Death represents to people the loss of the string of reality with loss of time connection.

If we could have total assurance that we would continue to exist in some form after death, then would we fear death? Just as the light of creation persists as a spark in each of us, so the darkness of the fear of the unknown also resides. Yet the key to creativity is the ability to face the great unknown that we must face before we take the creative step and leave behind our tags, prejudices, strings—and logic. Logic is not a tool for creativity. It has value only in retrospect.

In accessing light, we have to give up ordinary thinking that we use to approach daily life. We need to ready ourselves for new paradigms. We live from one thoughtform to the next. We have created these ready-made forms over long experience. Reality fits together for us in a link, one link fitting into the next, though we can only see those links in retrospect.

Realizing this, we can see the fright that many older adults develop when they lose their recent memories as in the Alzheimer's process, and their chain of reality no longer fits together because of memory gaps. Frightening.

When we cannot call up old thoughtforms to deal with matters in life, and we face the need of forming new approaches, we feel irritation. We argue. We feel resentment. We prejudge every situation in life and have prepared ready-made thoughtforms. We all have prejudice.

As the Hindus say, to enable creativity, there must also be destruction of the old forms. We must destroy old thoughtforms and prejudices. To do that, we can insert new conceptions, rather than try to suppress the old thinking. Suppression does not work. In the following sections I am going to apply what I have been saying about discrimination, identification, and understanding to common problems regarding conflict. It is pretty straight forward stuff and practical, and I mostly use couples as examples, as troubles in couples is something we all know something about. It may sound like a couples' therapy session, and in many ways, it is. You may already know everything I am about to say. You'll see very soon.

Argument

Conflict comes from our desire not to give up preconceptions and firmly-held thoughtforms. We cover that with such affirmations as being a man, being steady as a rock, being basically stubborn. What is the difference between stubbornness and having resolve? Stubbornness is inertia, resistance from being moved. Resolve is the necessary will to accomplish something and overcome fears. Will is necessary for growth as a person; stubbornness is the will *not* to grow, and it always has fear at its base. Resolve can cut through that fear.

When friction occurs between two people, I can guarantee both people have something to learn. You may not agree with that, but I very much believe it. In an argument the so-called winner is always wrong, unless you define winning as learning. Not many people do. We can win by resolving arguments. To resolve an argument, we first must understand the dynamics of argument. Figure (17) is the graph of any argument. It is from my book, *When Mirrors Become Windows*, which is, at the moment, unavailable.

Figure (17) Graph of Argument

You will note on the graph that argument begins with poor communication. Let us say two people are responding to one another through their well-worn forms of thought. Tension begins to build, and at first, they feel this as loneliness and estrangement. Some people stop at this point on the argument curve and stay that way. They carry forward a state of loneliness by maintaining poor communication, and they appear to others as unreal. That is why it is so difficult to help a lonely person. Very few people want to be around an unpleasant lonely person.

More likely however, feelings begin to rise to the surface and friction results. Anger then appears, then fight or anger to hurt. Couples know the right buttons to push to hurt one another. Men, especially, do not allow themselves to go too far beyond the anger phase and into the anger to hurt stage, because they fear if they totally release their anger, they could not control it. They remain begrudgingly passive. Contact sports, wars, violent movies or games help to defuse it, but such things serve only as Tylenol to a problem that does not resolve. The manufactures of ulcer medication make billions off such people. It is a recurrent argument that does it. The bottom line is that someone is being hurt in the fracas, remembering that all anger is psychic hurt misdirected into anger. Everyone is familiar with the anger to hurt stage in argument. Some likely examples from a large repertoire:

"I hate being around you with other people."

"I've always hated (insert hated trait) in you."

"You're so stupid, just like your father (or mother)."

"You talk too much."

Suddenly and after having the same argument many times, the focus shifts.

"I can't stand it anymore!"

All at once, the person faces the blackness of the unknown, without the tyranny of previous thoughtforms. For an instant: no old baggage. At that point, you can see if the person has any courage to face the blackness of the unknown and make a decision to go forward.

At this crisis point, the argument combines risk and leap of faith into the category of *Conversion Process*. Conversion process is making a sudden decision to allow oneself to lose control (of previous thoughtforms) and go into the blackness of uncertainty, only to emerge into the realm of a new realization. That's a process in common with all growth and creativity. When the step is small, we call it learning. When the step is more inclusive, we call it understanding. When it is all-inclusive, we call it conversion. Intrinsic for allowing a growth experience, is taking a risk or 'giving up' which we see in all walks of life, be it religious, creative art, or even mountain climbing. It is that giving up, facing the blackness of the unknown, that allows the light to express: creativity cued by soul energy. Without that creative step, the argument will not proceed to resolution.

A decision for change is declared. Some possibilities:

"I want a divorce."

"Take this job and shove it."

"See how you like getting along without me!" The husband (or wife) bolts the room and slams the door. Takes off in the car.

Something is thrown and destroyed, usually with symbolic significance, such as a photo in a frame or something previously valued by one of the arguers.

Hopefully no one is hit, or worse.

The point I am trying to make is that if things deteriorate enough for an argument to occur, and it is a recurring argument. Then in the resolution of it, and in order for the argument to proceed to resolution, there will be psychic hurt. I am NOT referring to disagreements, only recurring arguments. Maybe very emotionally laden arguments are always recurring arguments. I do not know that. Regardless, with civilized people, the hurt should only be emotional. But again, all anger is hurt.

One of the arguers might say, after his/her partner breaks down in sobs from the hurt, "Oh, what have I done!" realizing his/her participation in the argument. Or the person feels that the relationship really is not ending, but it needs adjustment. Or the person sees the situation from a new perspective and identifies the problems, realizing that she no longer wants to participate in this failed relationship. Or the one dashing out the door, cools off in the car, considers improving things and returns. Or the person keeps driving and returns angry, and somewhat later climbs into the narcotic of someone else's arms. That's not creative, nor will growth ensue. That's the subject for novels and reality TV.

It is worth remembering that without risk, there will be no resolution of recurring argument. And again, that risk is abandoning treasured and long-held thoughtforms that prevent change. We cannot know how we feel about a decision until after we have taken that decision. Then we can see how we feel. Up to then, all is speculation. We fear making the decision that we think we *have* to make in order to solve the problem that the argument is supposedly about, for instance, the only way forward is thought to be divorce. If that's true, then that fear has to be faced, and the uncertainty of the future, felt. If it's: "No, I don't want to leave. We have to find a solution." Then counselling is a good way to find solutions. Counsellors (hopefully) are trained in helping people find solutions.

One thing I clearly know: If one of the partners escapes into the arms of another, then there will be no growth whatsoever. The escaping partner feels the sudden exhilaration of engaging in a relationship without the years of old emotional baggage attached. Of course, the sex is fantastic. Of course, they are perfectly compatible. Of course, it is fantasy. The unresolved argument persists, because both people in the argument have something important to learn. No growth results. The one left behind experiences hurt, and that hurt can engender growth, unless the person just hangs on to anger and wants revenge. Same with alcohol or drugs. If one or both of the partners resorts to chemicals when the going gets rough, then no growth will ever occur. When you avoid hurt through chemicals, then you avoid growth. I will discuss hurt much more in later chapters.

To sum up, the reason for the argument graph is to show

that all arguments have to resolve only through learning on both sides. Both. That learning requires dropping the long-loved and endlessly thought-about thoughtforms. You would have the task of identifying your own thoughtforms which reside at the base of the argument. Dropping that thoughtform is a leap of faith, and when whatever you feared does *not* take place, then learning occurs, and if both go through that same process, the argument is permanently over; and both you would march onward to further learning. Sounds like a pain. And it is. But both partners in any relationship must grow together, or the partnership dissolves, or continues with all the desperation of hell. No way to live. Take your choice.

Argument and Sacrifice

Taking risks is risky. Hurting anyone physically (as a risk-taking maneuver) is against the rules, though used frequently in the ritual called *sacrifice*. The ritual of sacrifice is a symbolic, or real, hurt used in an attempt to resolve an argument, even though it is a cop-out. Institutions and governments, with malice and forethought, use the ritual of sacrifice all the time. They fire the chief of staff, or the teacher, or the head of whatever, to make everyone think that they have found the cause of the trouble. Thereafter the government, or the institution continues to do what it has done all along, only to discover later that the person (or thing destroyed) turns out to be someone or something of critical importance, unrealized before. With that loss and the subsequent turmoil, the argument has resolution.

That which we pick to destroy in the ritual of sacrifice always has value as a symbol. For instance, the wife throws a picture at the husband, barely missing him. The picture that shatters against the wall turns out to be of his mother. The couple later reaches the conclusion, after learning takes place, that the problems involved an interfering mother . . . and a complacent son.

When governments sacrifice real people, and after learning takes place years later, people build myths surrounding such people. Martyrdom remains the ultimate method of sacrifice, in this case, self-sacrifice. The martyr uses self-sacrifice to push an argument to resolution.

Suicide gestures also come under the heading of the ritual of sacrifice. The light cutting of the wrists or, maybe, taking 6 Tylenol and 10 Midol tablets, or some such as a gesture to suicide demonstrates the desperate attempt at self-sacrifice with hurt in an attempt to resolve an argument. With nearly everyone having dangerous drugs on hand, suicide gestures have taken on a new dimension.

As a method, argument is second best to having enough mental awareness to spot the things that trigger arguments and take care of it on the spot. We then resist the temptation to take the next easy step and fling ourselves into an emotional thoughtform and resort to argument. We know from having mentally marked the issue that we and the other person have something to learn.

Through will (or resolve) we set out to discover what it is that we both have to learn. This is called the middle road, and that road begins by our realizing that the road connects to the light of our soul, and if we are going to have communication on this road, then we have to make contact with the other person's road. We learn where the other person is coming from, and we do not have to agree with her/his perspective. In fact, we never do at first. How could we, since we had not started out that way.

When we see the light in the other person, then we can make identification, and at that time we feel joy. Energy then flows between us, and we both grow, but most certainly in our own ways. The manifestation of that process we call love, or in the Christian terms, manifesting the love of Christ, and in some sects of Judaism, manifesting the Divine Spark. Muslims have the concept of brotherhood. Muslims and Jews have much the same philosophical bulwarks since they trace their origins to Abraham and his two sons, so the Bible contends. Being so similar, maybe that is why they don't get along well. You see that in couples who are very much alike, and they argue a lot; and any differences, no matter how small, become a big deal. On the other hand, complementary couples become assets to one other.

Bargaining

The concept of the middle road, and how we can follow the light to resolve argument also works in a process called Bargaining. Bargaining bases itself on the firm but hard to accept rule:

If there is a problem that produces friction, the problem resides with both partners in the conflict 50/50.

Looking back on the process of argument, you remember that both partners in the conflict have something to learn. Bargaining is no different, and we all know that in conflict, whether we resort to arguing or not, each individual involved thinks that she/he knows what is right, and the other is wrong. I am NOT talking about disagreements over facts, like, are we on the wrong road and lost? No. I *am* talking about, "You always think you know the right way, and you don't!" Or something like that. You know what I mean. It's deeper than a disagreement.

Bargaining is a more mechanical means to an end, and it requires taking the above 50/50 rule on faith, at first. The realization of its truth opens the way for resolution of the conflict through a road that neither partner has traveled before. On that road, both learn, and both learn together, which turns out as a process of love. Bargaining merely sets the table for inner change, if adhered to. It is so simple, it hardly needs explanation. It is for two people who find themselves in conflict, but they both want to solve the problem and get on with their lives and be happier, just as it was at the beginning.

You and the person with whom you have the conflict must look at each of the situations that produce recurrent friction. Each of you will have differing opinions as to cause. Cause does not matter. Fault does not matter. On separate piece of paper each writes what she/he thinks the other can do to alleviate the friction. Do this without consulting the other. Number those items. Each of you will end up with a list of things you think the *other* can do to reduce the friction.

Now the bargaining begins. You will trade items or groups of items: "I will do that, if you will do this," etc. You will find that you nicely bargain of all the items on each list. Thereafter, each has to keep the bargains. The lists can be revised later, if both agree, as conditions change—which they will. Just keeping the bargains will raise tensions, and problems will surface; but by maintaining this middle road, the conditions encourage a perspective different enough that learning and change can occur. When the going gets rough, you have to return to the original 50/50 ground rule. That's a hard pill to swallow, but a necessary one and a valid one.

It has value to remember that the things we trade in bargaining are things we want the *other* person to do. Ultimately, what we want is for the other person to do what she/he wants to do: "I want you to do what *you* want." Life then becomes both of you doing what you want to do (along with things you know you *have* to do). The magic (and adventure) is that you get to do it together, and hopefully have fun doing it. Any of what I am saying is out the window if there is drug addiction or dependence.

All of this takes place in everyday life, and nearly none of us is aware how often we quickly make bargains to reduce friction, so quickly that we do not remember doing it. Friction begins if we are made to doubt ourselves and our thoughtforms. Let me give you an example from my own life.

My wife and I had at dinner with friends at our house. She handles the food. I handle all the infrastructure involved. She was telling me to do something, like, Phil, go do such and such. I turned to her and said, "You know, you could say, please." (At that moment, deep inside, I was having the feeling that I did not want to appear as my wife's servant.)

What I said queued a spite of anger from her. "Well, *you* could say, 'Thank you' more."

"Okay", I said, "I will say, thank you, more if you will say, please more."

She agreed. Bargain made. The revealing thing is that when we are very busy in social situations, as we have many people over, we don't always remember to adhere to the bargain. But it does not matter anymore. The affective (emotional) side of the thoughtform that caused the friction is no longer present, and the whole issue no longer matters.

In the case of argument, one jumps into the blackness of the unknown to reach a different perspective. In bargaining, the ground rules force a different perspective. Even with self-doubt, the doubt itself gives us another perspective, and with the consequent stereoscopic inner vision, we can go forth. Both Argument and Bargaining can lead to the path of light. Once learned, bargaining can be applied to all relationships. Discrimination holds the key--identifying what you are thinking. Personally, I think that bargaining should be part of the wedding vows, as bargaining is essential to any happy marriage.

Concepts to Carry Over

Matter and energy remain in equilibrium. Light indicates intensity of reaction. We use light to foundation ourselves. Light is a portion of all light, a quantum of which (the light of creation) remains with us. The soul contains light (light of the soul) and matter (the soul body). The light of the soul represents our finest potential.

Our thoughts connect to emotions, called thoughtforms. We surround ourselves by these thought/emotional energies from multiple sources. To discriminate these energies, we first must integrate our personalities.

An energy field surrounds each of us, and it is the first to react to incoming energies. To see the light in others (a process of discrimination) we use the methods of identification in which we drop our fears and open ourselves without prejudgments.

Seeing the light and solving arguments is a dialectic in which Conversion Process plays a key role and involves a leap of faith which drops preexisting thoughtforms. Sacrifice is frequently used in argument resolution. The process of Bargaining, through its particular ground rules, frames growth through understanding.

TIME

The instantaneous present is the door to the eternal. What does that mean, and is it true? Let me attempt to answer both of those questions. To do so I have a great deal of material to cover, and much of it is abstract.

We know the present, we think. Psychologists tell us that we must not have our minds on the past or the future, but right on the present. We used the phrase, "Right on," in the 1960s. What is the present? Does the present exist? If it does, do we ever experience it?

The present represents the hypothetical line that divides the future from the past. Does the future merely connect with the past without an instant that we could call the present? We doubt that. We have yet to experience the future, and we have already experienced the past. There must exist an interval between the two. We shall assume so, for the time being.

A Creature Watching

Let us say a creature exists that could watch us from afar and could see our past and could also see where we are heading in the future. To the creature, we would appear as little organisms generating lines of experience. We experience our own reality, and if something were watching us, that creature would see much more than what we were seeing. Even in our own surroundings, we miss much. I can imagine that we would appear to this creature as I would view an ant traveling its own path, totally unaware of my shoe located right next to it. The creature has greater consciousness compared to both the ant and me. Consciousness is relative.

The ant builds its nest and does certain activities at certain times of the year, but we would have to suppose that the ant is at the present all the time. It does not have enough of a mind or brain to have preoccupation with the past or concern for the future. What it *does* have is what we call instinct. All that exists for the ant is the instantaneous interval connecting the future and the past. Little brain, little strain. Being only at the present, time as we know it, cannot exist, because the future and the past do not exist conceptually for the ant.

For us humans there exist not a point of time but a line of time, a succession of events that we can time with our clocks. That is our time, and our time does not relate to the ant's sense of time, which does not exist, or if it does, it would be much different that our time. *Time, then, may link to level of consciousness.*

How about time as it relates to the creature who is watching us? As the creature watches us, we would go about our business, thinking about what we are doing, mostly lost in our activities; and we could be looking at the clock while we were doing so. Since the creature can see our past and pretty much can tell what we are heading for in the future, then the creature's perception of time must differ from ours. The creature has *expanded* consciousness compared to us, and the creature would have a different time line than we have. (Note: the degree of consciousness still holds as degree of awareness that we all are One, though a bit convoluted in this example.).

Let us say the creature decides to intervene in our life and make itself known to us. The first difficulty would be the time dimensional difference, and the creature could not intervene unless it came onto our line of time. We, in turn, could not contact the creature unless we entered its time realm.

"Ha!" someone might say, "The creature and you cannot communicate unless you enter the its time, yet by analogy, you can intervene into the life of the ant without changing time lines. You can stomp on the ant with your foot. What's the difference?"

It's true. I don't have to change my awareness of time in order to make a change in the life of the ant.

Stomp!

I just crushed the ant. As I do, that action must affect my life in some way. Whatever I do to kill the ant or intervene in the ant's life in *any* way will create evidence of this intervention in my own life in some big or very little way—the ant's juice sticks to my shoe, and I can smell it. Perhaps, I missed the ant with my shoe, and the ant stings me, and I die of an allergic reaction. Big or little, we can find evidence of our encounter, somewhere.

I cannot change the ant's life without changing mine. During that contact, we must reside at the same time, even for that very instant, and there must be an act of will on my part for this action. I have to do something to make changes in the ant's life.

We identify *will* on my part, and we can see that by this act of will the time line of experience of myself and the ant intersect. To the ant my actions must be on a scale of greatness and my motives so unexplainable that I seem like a god. Just the same with ourselves and the creature watching us. In order to intervene in our lives, the creature has to have the will to intervene and then to make the action, and that action would affect the creature itself in some way, just as my action with the ant had to affect the ant and myself in some way.

If the creature intervenes, then there will also be a point of intersection of our time sequences, if only during that action. The perception of the sequence of time is different in each of us, but when we interact, our lines of time intersect. Let us explore these ideas a bit further with an example in our own line of time as humans.

Let us say that I am a trained observer in an experiment, and I am watching you through a one-way mirror while you do something very interesting to you. You are totally involved in your task, and I must fill out a form on your behavior while you are doing the task. What will be the *perceptions* of time held by you and by me?

At the end of the task, someone could ask you how long you were working at the task. If you enjoyed the task, you would likely say that it seemed like a short time. I, on the other hand, having to fill out a boring data sheet on your activities, would say that the time seemed very long.

What if we were out in space and had no sun to go by, and our clocks had no relevance (as our clocks base themselves on the sun). Both you and I would have two separate times. Does that mean because I felt the time was longer that I actually aged biologically more than you during that time, just because the time seemed longer to me?

No, someone might reply. The aging process has its own time clocks, biological clocks, and we can go by these. Is that true? Well, no, not exactly. People age at different rates. Some people are old at fifty. Others are young at fifty. Does the perception of time have any relationship to how fast a person ages biologically? That, unfortunately, is an unanswered question.

The ant lives a shorter life than we. Does grade of consciousness have anything to say about life span? If I use analogy, then I would say that the creature watching me would likely have a longer life than mine, but that amount of time would exist on our clocks. In the creature's own time frame, its life span would have equivalence to other creatures living in that time perspective. Let us leave the ant for a while and proceed by examining those moments when we intersect another's line of time.

Direction

Within any time frame, we tend to go on a track. A track or a line has only one dimension, and we go from the past to the future along this track. Intervals of time proceed as we experience our individual tracks, and when we come into contact with another person our tracks intersect. During a conversation we construct a more or less mutual track between ourselves and whomever we are speaking with, and the direction of the conversation (the direction of the track) varies depending on how self-centered or other directed we are.

Once the conversation stops, and we separate, we either climb back onto our usual track direction; or else our life has changed somewhat by the encounter, and a new *direction* assumes. Rare, unfortunately, is the situation where people converse and both proceed in new directions due to the encounter. We might call that a *creative conversation*.

As we travel on our tracks, we use the equation $D = R \times T$. Distance equals the rate multiplied by time. The distance varies according to how fast we are going and the amount of time that has elapsed. What does that have to do with direction?

Direction is what we were talking about at the beginning of any conversation, when the tracks of time intersect. Before that time our tracks were proceeding in most any direction. Direction relates to quality and content and requires a goal, either known or unrealized. Direction means how direct or focused we are approaching that goal. Are our tracks mere wanderings in the sand, or do we see a straight, determined path? When we meet one another and start a conversation, we intersect, and then we begin from a common point for just an instant.

Direction becomes critical for the significance and continuance of the conversation. You have found in life that with very good friends, even if you have not seen them for a long time, you pick up where you left off, and the time did not matter. You begin to travel on a common track and find new directions for both. This is an exciting time of creative conversation.

When you leave your friend, you feel different, so does the friend, and you find that each has changed. The tracks remain close to parallel. This is not to say that the character of the lives of each is the same. No. The lives can differ as difference can be. The importance is direction, something that we identify with soul values, and the direction remains independent of life style, vocation, and personality.

Interesting to note that during the time spent with close friends, the perception of time compresses such that time flies by. Perceptions of time compress when we find new directions. Perceptions of time expand or lengthen when we plod along in the same direction for any length of time. "It was a long day," remarks a person who has had to travel a path of habit and structure, dealing with obstacles along the way but continuing in that same direction. This is work. If, however, the person had the power to deal with the obstacles in a creative fashion, then time compresses. Having to live on the habitual path and struggle with it, always in the same direction, causes stress. Time seems long, and the stress shortens lifespan.

Within the directions that we travel on our own separate tracks we feel affirmed by our friends. As they change, many times we feel involved in those changes. We might say to ourselves, "He seemed to change after I talked with him." Not to admit that we both changed in the process, because change in one always involves change in the other, though we are mostly too distracted to realize it.

Time Orders Our Lives

In our own lives, time perception has importance. The actual time related to the phases of the sun as it crosses our sky has little relevance. The time segments with which we divide our lives by use of the clock merely order our lives and allow us to keep commitments, a process that tends to keep us going in the same direction.

We allow the clock to shorten or lengthen whatever we happen to be doing, no matter how interesting we might find the activity. The clock pulls us back on track. This is part of the so-called socialization of children. They learn to tell time, and they reluctantly abide by the clock rather than allow interest alone to determine how long any particular activity rates for time spent. Native Americans still puzzle at what clocks do to people. I, as a physician, must treat diseases potentiated by the strict divisions of time.

Time Dimensions

Let us say that we live on a line, a line of time. We can see behind us and say that an infinite amount of time extends backwards from where we stand now. We could look forward and say that an endless amount of time extends forward, to the future. We do not know the bounds of the future or the past. We call it endless: infinite.

Let us say that our line of time, which only has the *one dimension,* that of length, begins to move. I mean our whole line begins to move in a direction, say sideways, not that the line just changes direction, but our entire line of time, past and future, begins to move in a direction. Our realm of experience then would change from one of length, living on a linear time line, to one of breath, also. From geometry, we know that as an entire line moves sidewise, it makes a plane or a flat surface, something that now has *two dimensions*, Figure (18).

What would that do to us? For a start, it would completely transform what we had thought was infinity into a new infinity: from an endless line to an endless flat plane. We would have much more freedom, on a plane, than we have on a line of time. We would now have a plane to deal with, and linear time would become mostly obsolete, but still useful. What would that be like?

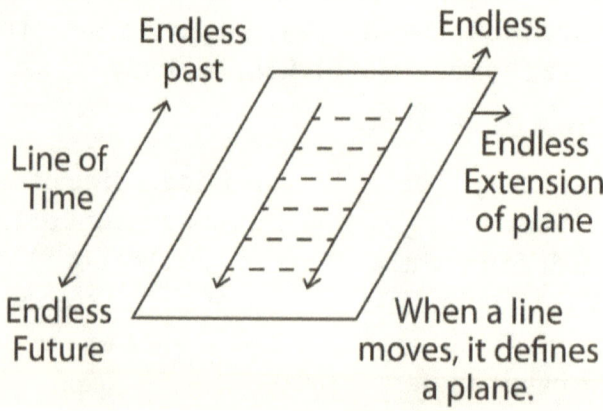

Figure (18) Two Dimensions of Time

Time would still exist, but not having experienced such a transformation, we would not have the words to describe the experience. It would transform time as we know it. It would be a sort of apocalypse.

Would that end time? Hardly. That plane of time can expand again by moving in space, much as the line did previously, but by the planar surface moving it creates a *three-dimensional* world of time, Figure (19).

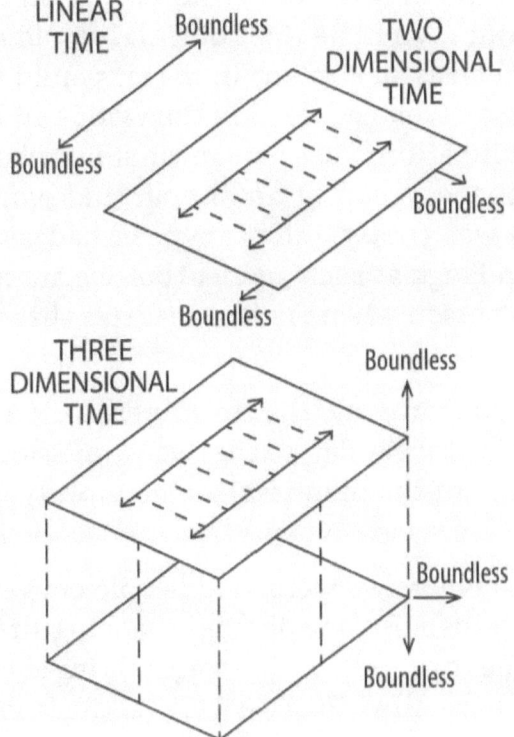

Figure (19) Three Dimensions of Time

As a plane moves it makes a three-dimensional container or space. The container described--an endless three-dimensional space without boundaries--would define a new infinity, yet containing the infinities of both linear time and its extension into two-dimensional time.

Do all these expansions of consciousness mean that the infinity on our little time line as we live our lives, is really not infinite? No. Our line can still extend to infinity, whatever that is. It does mean that our infinity exists only as a part of greater infinities, infinities that include realms of which we would only see indications, and even when we do see these indications, we would tend to ignore them, or misinterpret them, because we have few words to label these encounters. Jung called their intersections, synchronicities.

Events coming from another dimension of infinity, if they interfered with our own little lines of time, would appear as magic to us, much as an airplane in the air would appear as magic to a person who had lived in the woods all his life and never seen civilization. That person would not have words to name the phenomenon of an aircraft and would likely maintain that it was a noisy bird of a type he had never seen before. We would laugh at such an idea, but we are not much ahead of such a person when we face realities that we cannot explain.

How might we go from, say, the linear perception of time, the type that we lead every day, to the two-dimensional perception of time, that of the infinite plane previously described?

Let us go back to where we were, two people conversing, and our linear tracks bending and perhaps changing direction during and after the conversation. We were talking about changing direction, but the directions we were talking about involved one dimensional lines.

A line of time has only one dimension, that of length. When approached by another line of time (another individual), and the lines make contact, such as in a conversation, the lines intersect. At that moment of contact, we have an instant of pause. We have an interlude, a pause between two activities, a critical juncture. We may join that person's time line for a while, but once we are out of contact, we go separate directions. We may be going in other directions during the entire conversation, and the content of the conversation may have no meaning to us. It is significant that at the instant of the beginning of the encounter, the lines do intersect, and we are at least momentarily on each other's time line.

Polarity

Let us go deeper into this subject, as the answer to our questions may reside in something more abstract. A line has length, as does time as we perceive it. If we describe any amount of time, then we arbitrarily divide off a segment of the line of time. By so doing, we polarize life.

In infinity there exists no polarity—no start, no finish, no negative, no positive, no yes, no no. Once we make divisions into linear time, then we must also have polarity. Our existence in time implies polarity. If we want to stretch our imagination to infinity, then we must answer the question: When did matter (the universe) begin? Has it existed all along, with infinity? Or did it come about on some segment of the time line?

If we believe in creation of some sort, then we have to accept the fact that polarity, of all types, comes about when something is created. In Eastern thought they use the term duality. Everything is dual. But, they would say, the ultimate reality, apart from the universe, is One. We could get into all sorts of discussion of God transcendent and God immanent, but those issues have no importance at this point.

We come back to polarity. The history of the world has been one largely of the resolution of polarity. Polarity means difference, and the problems of the world result from differences, and we seek to resolve these differences through a variety of means, mostly by some sort of force.

The more contact we have with one another, the greater opportunity we have to experience, at least briefly, the time line of another. Such contact does not resolve the inherent polarity built into linear time. Each life on individual tracks traveling at different rates of perception and in different directions, but all on a flat plane—myriads of lines on a flat plane. Scribbles?

Moving Between Time Dimensions: Compassion

One way we have to experience the second dimension of time, not to mention that time has many more dimensions than that, is to jump into another's skin, as has been before, and thereby experience time as does that person. Such an experience, called *identification,* allows us to extend ourselves across the plane of time, giving us a comparison—of their time and our time—and thereby describe a plane instead of a one-dimensional line. We describe a plane either by moving our own line sideways or by intersecting another's line.

Even for an instant, to see as does another, or perceive as does another, or think as does another, we call that form of identification *compassion*. We cannot *be* another, but we can put ourselves in the place of another and attempt to understand what they experience. Compassion occurs at two different levels: (1) *I understand what you are saying,* and (2) *I can feel what you feel.* I cannot be you. I can have compassion by having experienced similar situations as you, either at an emotional level or at an understanding level, or both.

In Eastern thought, an old soul is one that has had many incarnations and consequently had many experiences. Learning takes place in each incarnation, and so an old soul is a great soul because of its capacity to generate a personality with great compassion. Jesus, according to the Christians, would serve as an example, with Godlike powers of compassion. The Christian believes that she/he derives the positive aspects of compassion from the compassionate energy of Jesus, as Christ.

Through compassion, we move our entire time line. Therefore, if we want to experience the expansion of consciousness inherent in the second dimension of time through compassion, then we must do our thing in relation to someone else's thing. The writer and philosopher, Sartre, once said that hell was other people. Perhaps to him that was true, but it is equally true that the key to the second dimension of time, call it the soul world, or heaven, or whatever we want, that second dimension depends on other people; and we must exercise compassion in order to access it.

The word "exercise" in relation to compassion is proper. We exercise compassion just as we would exercise anything else. Through exercise and practice comes expertise. Through experience comes expertise, as long as we learn from experience and do not continually make the same mistakes over and over again. It is at the moment of contact, the moment of intersection of two-time lines, the instant of direct contact, that we make the decision whether or not to travel on another's line, and how completely. Maybe we have decided to entice the other person to travel in our direction? That's the work of a salesman, or a demagogue.

With the exercise of compassion, we make the instantaneous decision to travel in conjunction with the other person. It so happens that if the other person also has made that decision at that same time, then we feel that we have found a brother or sister, and we both willingly travel together on what turns out as a creative conversation. The creativity in the conversation is that both exercise the will to stay together, and neither knows where that conversation will lead. Creative conversations lead us in directions that neither participant would have gone alone. That's fun.

What about the case where we decide to stay on the time line of another, and the other person does not make such a decision? The other person is only concerned about self. That becomes the exercise part of compassion. We must exercise the muscle of the will to stay with that person, and we must apply the will to search inside the other to find similar feelings or similar thoughts, and then identify with that person—find the light. Once we make that identification, then we may decide to hold there and allow the person to find him or herself, or we might make the decision to intervene in the other person's life. This last decision has difficulties.

In order to intervene, we cannot suggest something that comes from our own system. To avoid problems, we must choose something within the system of the other person, and perhaps one step forward. We take a chance to even suggest that step unless the other person asks for help, and even then, we must exercise great caution. We *can* shine light on the path, helping the other person see the next step for him or herself. We call that *service*.

Destination

I was speaking of directions of time lines, and those directions inherently involve the second dimension of time. Let us look more at direction.

If we return to the simple equation $D = R \times T$, Distance equals Rate multiplied by Time, and we transpose that equation to $T = D / R$, then Time equals Distance divided by Rate. Time, as we experience it on our individual lines of time, involves distance and rate—how far we have come and how fast. How far we have come toward what? Now we get to *direction*. When we speak of direction, we talk of destination.

If we did not believe in anything besides what we could sense with our five senses, and we do not include the mind as a sixth sense, then we would have to believe that the only destination is death. Not many people believe that death is a destination. Most people will accept that death is a stage or a period of transition.

Let us say that death is not the destination. If we do that, then we must consider that the different directions each of us takes on our individual time lines have some significance, and that there is a preferred direction, one better than another. Is that true? I have just maintained that in compassion we identify with another's time line, and that the most compassionate thing we could do is shine light on that person's next step. Next step toward where?

Is there a destination somewhere on this plane of time lines, the plane making up all the time lines of everyone, and each time line only consisting of a one-dimensional line, and each of us unaware of any other portion of that second dimension unless we begin to understand the line of another? Are we all heading for some goal on that plane, Figure (20)?

Figure 20 Goal of Existence

Let us use the microscopes of our minds to look a little closer. The second dimension of time is the plane of time on which all of our little time lines run. We can change the directions of our time lines to conform to the time lines of others. Even though we derive some experience of the second dimension of time through, say compassion, we still remain on our lines situated on the plane of time. We still live on the one-dimensional line.

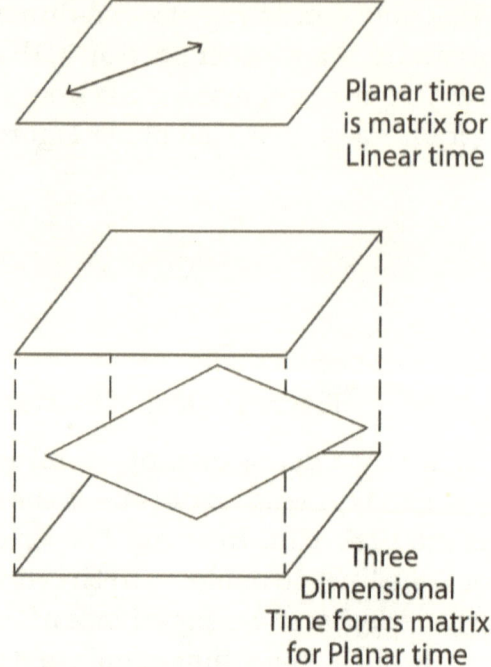

Figure (21) Matrices of Time

If we had the ability to live the plane of the second dimension of time, then we would traverse in our everyday lives a two-dimensional concept of time. Our area of perception would consist of the plane. The plane, as was said before, is only part of a third dimension of time. Whatever level of time we exist on, the next level up is the matrix in which our own time exists, Figure (21) Matrices of Time.

We can experience the second dimension of time when we exercise compassion, and if we become good enough at it, we can move at will from our time line to the time line of another. The exercise of compassion gives us the ability, or the muscle strength, or *will* sufficient to pull or drag our lines sidewise (efforts toward identification and understanding); and as we do, we describe a plane. While we are doing that, we experience what is involved in the second dimension of time. Movement or change serves as the key; *will* provides the horsepower. Do we have any other experiences to which we might compare this?

We know that in dreams we are not restricted to a line of time. The problem with dreams is that nearly all of us do not have conscious control during a dream. The dream carries us along as if our conscious 'I' had no say. Dreams have the concept of time of the second dimension, but they consist of projections of emotions and thoughts, worked out in symbols: the dream content. Dreams do have sequence as we have in our linear time, but we can jump that sequencing at whim.

It would be like the creature viewing us. The creature can see our sequences of time. The creature has a greater selection of ways to place sequence by jumping linear time, seeing what lies ahead and behind, much as I could view the ant running around next to my shoe.

What would be the relationship between the plane of time found in dreams and the concept of time that we might experience in compassion? Does that connect in any way to the possibility that there might exist a goal somewhere on our line of time? You may be a bit frustrated at this point as I seem to be bouncing about. Please have patience while I try to pull this all together. Thank you.

Plane of Mind as a Destination

First let us look at destination within the realms of lines of time. We would think if there were a destination, it would lead us to the second dimension of time. That dimension in itself is a destination. We would then have direct access to the second dimension of time and whatever powers and responsibilities that might go with it. I think we would do well to rename this plane of the second dimension of time, the *Plane of Mind*.

I say the Plane of Mind for several reasons. In the first-place mind can only exist in polarity. Mind comes about as a result of polarity due to experience. Within the polarity of our one-dimensional lines, we have the capability of seeking out life, much as a person would search in the dark with a flashlight. We have dependency on what we encounter. This is much as the animals live, except that we have more ability to adjust our environment than an animal.

In a human, life consists of more than just cuing on whatever we happen to encounter and dealing with it in a practical way. We have higher thoughts: abstract, creative, and intuitive thoughts. This knowledge comes not from direct encounters with the environment, but from within us. We cannot order up these thoughts as in the case of logical thought—one step leading to the next, linear, sequenced.

Perhaps the timelessness of higher thought gives us a hint that thoughts of such an abstract nature may have a relationship to the Plane of Mind, or the second dimension of time. We live and suffer in our ordered linear time, but our lines of time traverse the matrix of the next time level. Creativity tells us something of what goes on at that other level of time. Dreams give us hints on how sequencing of events can occur yet out of joint with time as we know it.

We look inward. The Plane of Mind exists in each of us, and all the individual planes of all people exist also in that same time framework simultaneously—individual and group planes, all along the same idea. That is a concept difficult to picture. We live within that Plane of Mind all the time, only we do not realize it, like the air we breathe. This plane must contain the source of our thought, or if not the ultimate source, then at least the repository of the thought at the basis for creativity in the form of ideas. Let us try to visualize this Plane of Mind in some way that might have meaning.

If I have my own Plane of Mind, then on that plane I have my own infinity; only I have not developed the ability in my lifetime to extend my thought to infinite thought. Instead I have a limited amount of this mental matter and a

THE RELIGION OF PHYSICS

limited ability to access ideas. I have accessed some of these thoughts during my lifetime, and these ideas remain part of me as abstract thoughts that participate in my own life. That thought material stands as the philosophy behind my everyday actions.

The ideas that foundation me as a person make up part of the Plane of Mind, but I am only part. The ideas behind the selves of other people make up the rest of the parts of this plane. We see something like a patchwork quilt, my participation being only one patch. The quilt of the Plane of Mind is all *one*, but we, as individuals, only participate in a section, surrounded and touching others at the same time.

If we conceive of this thought-patch-work-quilt as one of energy, with *concordant combinations* of thought grouped together, then the areas of energy groupings (the patches of the quilt) form larger and larger energy patterns throughout the quilt. Our separate patches do not group randomly but in patterns of related energy--soul brothers and sisters so to speak--and all of us together make up the Plane of Mind. Is that the goal, to realize that though we live in linear time, within each of us exists an abstract world of ideas that determines how we perceive experience?

We could call this patchwork quilt the idea level or even the *causal* level, as ideas precede any creative action on our part, the time mechanics being the same or similar to what we experience in dreams. We could also call this heaven. Why? Imagine this: You are surrounded by those you love and relate to best of all. Whatever you want, you can create through the power of thought. You can turn linear time around in any way you want. What a weekend away!

Experience will condition this mental level of existence. That means all the little lines of time marked out on the plane of the second dimension of planar time mark out the areas of human experience so far. We remember that an infinity exists for each level of time. At this Plane of Mind level, the infinite extension of the ideas behind human experience would extend to the ideas behind the infinite possibilities of human experience, if we can conceive of humans existing for an infinite time; and if they cannot, then just a long blank space extending to infinity after we are gone. Some would say that such is impossible. The universe was created for humans, and its existence and evolution depend on humans. Who will the observers be?

This very grand blueprint for human potential would extend into the eternal. Since this blueprint would include all human possibilities, we would have to call it a plan. All human potential resides in this plan. On this Plane of Mind, we saw only that amount of area traversed by us humans. There is much more to that plane with its potential extension into eternity. Since we are talking about the soul level of existence, and it consists at least of the collection of all souls, then we can call it *plan* and its eternal extension *Universal Soul.*

Plan and the Second Dimension of Time

At the plan level we have what we can call the eternal plan for human potentiality or perfection. Perfect as it is, it must have come from the same source as the universe. Therefore, it is no more perfect than that. Who said that the universe had to be perfect? It is not more perfect than its Creator, but since the universe is the only standard we have, we have to call it something. Suppose we say the Plan upholds the universal standard. We can put whatever value we want on that. All of this is still on the Plane of Mind. The eternal component represents only the highest portion of it, the Divine portion, if you wish. *The Divine portion of a dimension of time is its eternal aspect.*

We must realize that what I am calling the heaven level has limitations compared to the rest of the Divine portion or the broadest extension of the Plane of Mind. Remember that we are calling the eternal aspect, the *Universal Soul*. The individual soul consists of the individual portion of human potentiality. That which one human consists we could call the *causal* or the *soul body*. We can see that the so-called causal body would limit the experience that a human could have in the heaven realm, as it would have the same potential limitations that the human had during earthly existence.

Heaven

Heaven is nice, but it's limited. Too bad if we wanted to get to heaven to see the whole show. Eastern thought would say that we build our causal bodies through multiple incarnations, each time expanding them. When the causal body reaches a certain size or extent, it becomes a conscious aspect in our lives.

The only difference between Eastern thought and other thought concerning the heaven level (called Devan Chan in the East) is that Eastern thought sees the limited aspect of heaven, that it represents human potential during any particular incarnation. In Christian thought, heaven represents the end of the line unless the Christian accepts reincarnation.

A person existing at the soul level would exist within the matrix of the next level of time, *the third dimension of time*. Supposedly, we can consciously achieve that third level, but during most of soul existence the soul would have no more awareness of the existence of that third level than we would have of the Plane of Mind from the perspective of our linear time. The third dimension of time, is that our goal?

The 'I'

There are as many stages to infinity as there are levels of consciousness. The levels of consciousness depend on time, as we are beginning to realize. Consciousness also depends on the existence of an 'I.' If there is no 'I,' then we have no individuality, and we really do not exist as separate units. Where does this 'I' exist?

Within linear time, if all we see is what we feel we are, then we have defined that 'I.' That 'I' is on the physical level, and the 'I' makes decisions only based on what the 'I' perceives within physical (linear) existence. Let us call such an 'I', ego, small case e. The ego, in linear time, makes decisions based on physical happenings, emotions, and concrete thinking, thinking based on what goes on: survival, logic, but *not* anything abstract, intuitional, or creative. The ego is the center of experience, and it makes decisions and holds responsibility.

On this linear level we sneak around on the plane of the second dimension of time, searching with our flashlights, only seeing what we illumine with our lights. We miss a lot, but we can have some inkling of the second dimension in the following situations:

1) The moment that we intersect another line of time, as in an encounter. When two lines intersect, they describe a plane.

2) When we are able to move our line laterally through space by the use of our *will* in the process of compassion or hold ourselves on the line of another during an encounter by the use of will, thereby relating to another's needs. Healing energies would also come in here.

3) When we begin to think linearly, but through analogy, we inscribe a circle and come back to where we started. Such circular thinking shows the relationship of all aspects in the system that we were thinking about. We end where we started, and we would have had a journey in between.

The interesting thing about this sort of *circular thinking* is that when we describe a closed system, we describe a plane, mostly a circle. By enclosing a small bit of the plane, we become more aware of that plane and our relation to it, and we describe a point in the center, a point of symmetry that becomes a temporary 'I' from which we can have realizations, Figure (22). Instantaneously positioned at the point of symmetry, we can have a flash of an idea, at that moment only. Once we leave that point of thinking, then that potential 'I' on the planar level will disappear. Such a process also forms a stage in the science of meditation.

When we do any of the things on the above options list, we define a different Ego. We could say that this Ego is "higher" than little ego within linear time, and during the above types of thinking we are acting from Ego (Large E) residing momentarily, usually, on the Plane of Mind. If we had achieved the ability,

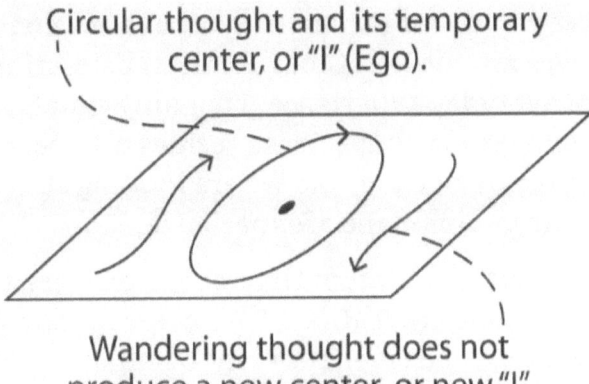

Figure (22) Circular Thinking

through practice, of staying at that level, then we would live it. Sooner or later, however, we would have to deal with the possibility that there resides another place for the Ego to sit, that of the third level of time: *third dimensional time*. Let us try to imagine what this third dimensional level of time might be like, or at least some of the characteristics of time on that level. Whoa. Are you feeling a bit over-whelmed? I feel that way, and I am only proof-reading it.

Third Dimensional Time

On the Plane of Mind, we live in energy, the energy of

thought. Time is like a dream. The energy of thought organizes into related energy groups. We each have an individual identity on that level, but the overall significance of the identity mixes with the identity of the group —we remember the patches and pattern on the quilt. All thinking at that level is abstract and forms the basis for ideas. The mental level is the abstraction of thought necessary for our existence in the next lower level, our linear level of everyday life.

If we wish to move into the third level of time, then we must assume that it serves as the necessary abstraction for the next level "below," that of the Plane of Mind. We are talking of abstracting the abstract. That's pretty abstract.

Let us pull out the microscope of our minds and turn up the power a few more notches. Further abstraction would lead us out of everyday experience. That only stands to reason, because our everyday experience resides on linear time on the first dimension of time, and we are looking at a plane totally removed from this plane of experience.

At this third level of time-consciousness, the only perception of the physical level, two levels down, would be through the second level, the Plane of Mind. The Plane of Mind is only an idea level and deals primarily with the abstract makeups of collections of individuals. An Ego existing on this *third* level would have access to these subjective groups. You could call that Ego a potential *Master* or super subjective leader to the individuals in the groups (the Leader within).

As we go from one level of time to another, we begin to think that a goal must exist down the line, somewhere. We likely do not have an endless series of dimensions of time. Somewhere there exists the origin of thought, thought unpolarized. We polarize thought through experience. Prior to experience, thought consists of only the mechanics of thought, the apparatus and system necessary for thought before given something to think about on the physical level. We could call that *non-polar thought or Divine thought,* if you like. It had to be the type of thinking prior to Creation, unless, of course, God had a friend, I mean, Friend. If you are an atheist, please do not be put off because I use the word Divine. By Divine, I mean only the extension to infinity of any process. I could call it cooked cabbage. What's the difference?

Assuming that higher levels do exist, and that the Ultimate level resides some way beyond, then we have to assume that these components of thought that arrive onto the Plane of Mind from the third level have pre-programming in some way. *Preprogrammed components of abstract thought.* What might they be?

Abstract Thought and the Third Dimension of Time

What are the components necessary to form an abstract thought? First, we would have to have an 'I' or some Ego to do the thinking. We have examined that before, and we will just assume that for the moment.

Will cannot exist as an isolated quality. The will has coloration or direction. Direction? What direction? We have optimism, and we can call that direction the Will-To-Good. Will needed a spin, and we have defined the most efficient as the good. In the chapter entitled, *E*, we saw how the movement of evolution toward the most efficient also defines the energy of love. *Therefore, we are saying that love and will mix together to give the beneficial spin to Will that we call the Will-To-Good that all religions define as part and parcel to their God (or Cooked cabbage to atheists).*

We are still examining the components to abstract thought, and we are aiming at the third level of time which we have said has the function of pre-programming the abstract thought that we find at the second dimension of time. So far, we have Will colored with the direction of Love. What else do we need?

We must also assume the power of thought itself. Thought had to percolate down into all creations. Nothing would exist without it. *Thought itself*, not what might be thought about, but thought itself, is the power to bring things together into concepts. Thought itself is much like a scaffolding, similar to a computer, where it took intelligence to form the system, but the system has no value other than through its applications. We can therefore see thought as a *tool*, the great tool of creation. (Mondrian comes to mind.)

We now have three components necessary for pre-programming the abstract thought that we see on the Plane of Mind, though this Plane of Mind gets the *material* for its thought from the physical level —everyday life, linear time. These three separate energies are love, thought, and will, the same components to consciousness that we have been talking about all along. Looking closer we can see that love is but an aspect of will, with will carrying the most power messages of the three energies. We can see thought as a mechanism, or a principle, or a scaffolding with which to create, formulate, and synthesize.

Thought, as a mechanism, can only create if it has something to work with, and it needs something like us and the physical world to do its job. It does not mean that only good things can be created with thought. The Will, from above, so to speak, imprints our universe with the positive spin of Good, a Divine good that we tag as the direction of evolution, also defined by me (and science) as the most efficient. This Good is beyond the good (and evil) that we identify in our everyday lives. We can call it the ultimate goodness of will, the *Will of God* if you wish. I am going to stop mentioning that cabbage bit.

We have the powerful energy of will (along with love) and that energy, through the scaffolding of thought, acts directly on the highest aspect of the second level that we have called the Plane of Mind. At the third level of time we have the ability to act directly only on the interrelated energy groupings of abstract thought on the second level, and not directly on the physical level, as that is two steps away. We cannot really label the divisions between the time dimensions *steps*. Each dimension impacts on the dimension down the line, but that action is one of abstracting, as we have said, just as our everyday activities have tracings leading to our philosophies or religious perceptions, leading to the abstractions involving our subjective group, and so on, and on.

Each level of time would include that which a lower level would have and more. Therefore, all levels have sequencing, which is characteristic to our level on linear time. At the Plane of Mind, we would have sequencing and—Is there a world for it?—especially since we do not have it in our experience. We have it in our dreams; so, let us look for a word that describes it. Let us call it *Abstract Time*, since that names the type of thought found on the Plane of Mind. We will say that on the second dimension of time, the Plane of Mind, we call the time component *abstract time*. In abstract time, we have sequencing and we have time out of joint. We can jump linear time either way, forward or backward, depending on our thought.

On the third level, we are abstracting the abstract, and we are using the three energies, actually two: will and love (together) plus thought. Thought is bare, only a mechanism, but power, nevertheless. The content of thought we derive in relation to our world of experience. At both the first and second levels of time, these powers have taken on content from their relationship with worldly experience. At this third level we have content, but it does not relate to what we have already experienced on the physical plane, but more so with what we can *potentially experience*.

The content of the third dimension of time expresses itself through our *potential*: the future. Yet that potential resides beyond the collection of our individual potentials as humans, and it comes closer to the potential of all of us as a group. This highly abstract collection of energies within the third dimension of time defines *Plan* and is not individual, but *Universal*. This is the blueprint Plan for our whole system, and it focuses on more than just humans, but humans have the central role (we believe, anyway). Taken altogether and to its *eternal aspect*, we can call this *Purpose*.

What would time and thought be like at this third level, the level of Purpose? To exist on this level would mean superhuman powers, powers that regular humans only have in a stepped down version. We would also exist two levels of time out of physical existence, and we could only participate in human existence at the very foundations of thought. If we could take on a human soul and a human body for a little while, we would really be something: A Son of God would be a good title for somebody like that. Can we purposively access this third level of time in any way as humans?

Spiral Thinking

The most abstract type of thinking that we use on the plane of *linear* time creates a center of symmetry and also a plane when we use circular thinking. We end where we begin. Such gives us the idea of interrelatedness, with the point of reference at the center of symmetry of the circle, and it allows us to tap into a small area of the plane of abstract thinking from the point of symmetry. New ideas suddenly appear, new to us anyway.

Let us say that we indulge in circular thinking, but instead of coming back to where we started, we learn something new. The learning increases our consciousness such that we can never return to the same starting place. We have grown. This type of thinking adds a *third dimension*. Let us call it spiral thinking, or evolutionary thinking. I consider this a form of meditation called *contemplation*.

Through contemplation, we create a center because the thought is circular with a location for an 'I' at the center, but we also add the dimension of depth that taps into the third dimension of time momentarily. At that moment, ideas can appear, but since these ideas are abstractions of the abstract groups of ideas (on the second level), we will receive a main idea followed by a flurry of related ideas—said to be a "Raincloud of Knowable Things," as the seed of the idea from the third level of time reveals its implications. You may remember that in the first chapter I gave a mechanistic explanation for this phenomenon, saying it was like reprogramming a computer. True enough, but let me return to this a bit later.

Spiral thinking leads to growth in concert with evolution. This growth means developing the powers of consciousness, and we know that changing consciousness involves changing time perceptions (the idea of ONE expands). Paramount in the process of change is the power of *will*, the most powerful of forces, and it comes already programmed with goodness. It needs a human mind to express it and to fulfill evolution, apparently. Such makes us humans the quite important vehicles for Divine (or spiritual) evolution on earth. What a responsibility! Unfortunately, we humans have trouble realizing that higher will as we usually clutter our minds with desire.

The Higher Will

Let us apply what we have been discussing to the problem of something as mundane as weight loss. Yes, weight loss, the growing scourge of our world.

"I got to lose ten pounds. I need to go on a diet!"

We usually fail. It seems as though we could have the necessary will to lose weight. Why can't we do something as simple as that? The reason is that we are still using the lower will. We try to suppress hunger and avoid the social aspects of eating, but we fight a losing battle.

Suppression only results in more resistance. We have inside us what we might call *little lives*, or the functions of our glands, cells, and bodily systems. All of these little lives have little minds of their own, not very developed, but they know what they want; and the force of all of them crying out for food is too much to resist. In addition, they resist even more when faced with suppression—just as you and I would. Such resistance is built into even the most primitive consciousness. Even atoms have such resistance.

The only way we can access the higher and more powerful will is through a higher cause. That type of thinking moves the plane of our thinking into a new dimension. If you had someone you loved and you had only a limited amount of food, you would likely give up your food for that loved one. You would lose weight!

The trouble with dieting, and the reason that it rarely results in long term weight loss is that there exists no higher cause. The usual reason for dieting is either we want to look better, feel better, or we want to live longer. Neither of these reasons will access any will power other than that associated with desire.

The body is the expert in desire. It has instinctual cravings, especially revolving around things that have something to do with survival such as hunger. If we can think of a way to relate reduced food intake to *spiritual* growth, or compassion, or love, then we will have solved our problem—that is, if we are also concerned about our weight.

The higher we place our consciousness and the consequent broader our frame of reference, the more power we access. We overeat, while others starve. Does that idea stop us from eating too much? No. Real concern about world hunger, the compassion for suffering people, and concern enough to do something about world hunger—all those things automatically broaden our perspective and raise our consciousness, giving us more (will) power. We could use that power in our daily lives with adhering to a proper diet.

The bottom line is that the solution to overweight is learning more about the power of the will and exercising it. The problem dieters face consists of increasing the horsepower of the will, but the real problem is that we cannot boost that power other than through *service*. First things first. Now if you will only buy my book on diets and include the daily vitamin supplements that I sell, then you will have the shape of the model I show on the cover. Hurry, before you gain more weight!

Joking aside, I am trying to define the higher will of the level of consciousness in the second and third dimensions of time. When we tap into them, then we have real (will) power. My only point is that the way to control ourselves is best approached through ways of increasing will power in general. Will *is* power. Thankfully, the unworthy cannot successfully access that power for extended lengths of time.

A few words about the unworthy. The power that the unworthy access for their own use, that power, though seemingly effective at the beginning, also focuses on the problems within the person exercising those powers, much like a cloud of dust catches a spotlight. Self-destruction eventually results. We can have some sense of security that good will result, eventually, in our world. The so-called evil people have something to fear in the power of will. It can make them, but it will destroy them as they obtain more power. The Hindus call the power of will, Siva, the Goddess of Destruction.

Will is immediate, and at the third level it would contain all the powers of consciousness that we can access. We call it by another name, *Spirit*. The will we see in our everyday life originates from Spirit, but when we use it as we do, the Spiritual aspect seems quite far away. I will call the third level of time, the *Level of Spirit*.

Will and the Third Dimensional Time

Thought and knowledge at this third level would be immediate (more akin to the speed of light squared) as they would be contained in Spirit. All the knowledge of the universe as we know it, or not know it, is there, because the universe is there. The point to be made is that all the knowledge of our universe is contained in this Will, all the secrets of nature. All knowledge already exists, if indeed the universe exists, because the creation of the universe required these very mechanisms contained in Will.

Knowledge is immediate at the third level of time. Precipitation of knowledge from the third level to the second level and thence to the first would occur like precipitation, rain, a rain of ideas as was pointed out earlier when the concept of spiral thinking was discussed. We can call that type of thinking *intuitive thinking, Pure Reason* (Kant), *or immediate thinking.*

You can pick whatever explanation you like for the Raincloud of Knowable Things, be it the mechanistic one I describe in the first chapter, or this one which uses dimensions of time to explain it. Either works. Nothing happens without the power of (good) will.

Back to the subject at hand, how might we perceive time on the third level? With no direct relation to experience as we know it on the physical level, time becomes the mathematical program of evolution. Time becomes Purpose. *Purpose becomes the depth in time, the third dimension of time, the dimension added by the spiral.*

Purpose to us is immediate. Yet on this third level, there would still exist linear time and sequencing, but we would have to exist on that level to appreciate those factors. All we can perceive at that level is purpose, which reflects that mathematical or symbolic program for evolution, called Spirit, but considerably stepped down in power at our level of existence. Suffice to say, we do not understand Purpose. *Evolution as we define it does not exist at the third dimension of time. Perhaps that is the veiled idea bolstering the notion that evolution does not exist.* We will talk more of this later.

The Creature, Redux

Let us go back to the beginning of this chapter and bring back the creature that was watching us. Let us say that this was a creature from the third dimension of time. Little that went on in our everyday lives would have enough significance to even cause ripples at the third level. Significant growth would cause ripples, sort of like ripples in gravitational waves. If growth of consciousness (we could also call it spiritual growth) were significant enough, a human might become a disciple of that creature, even if the human did not realize it. A disciple would be a human on the physical level that had enough consciousness to participate more directly in the purpose of the third level. The creature becomes our potential super-human self, our Master within.

The creature is human!

THE CREATURE IS US.

Door to Eternity

Do we now have the necessary stuff behind us to answer the question posed at the very beginning of this essay: What does it mean when someone says that the door to the eternal is through the present? Let us see.

To have a present, we need an individual. An individual is a point of reference, an 'I-ness,' an ego. On the physical level we individuals move along tracing a line of time, but we really are only points with no other dimension. Remember, we are talking only of the point of reference called 'I.' With time, an 'I' collects experience, such experience is in reference to a time dimension.

We are dimensionless at our core and have no dimension except in relation to time, and time appears relative to what level of consciousness we are talking about. When we talk about the eternal, then we know that is relative, too. We also know that the level of consciousness can change. This dimensionless point of individuality, the 'I' --the focus of consciousness--can exist on other planes of time. The eternal on each of these levels of time is the *highest* or 'Divine' level of such realms, representing the infinite extension of that realm.

The individual, a dimensionless point, can only experience the present. By using the mechanism of mind, the individual can reflect about the future and reflect about the past. In these endeavors, the brain is quite useful. For experiencing the present, the individual has a body, and the body has sensory organs.

The present is the only possible existence for the 'I.' The 'I' collects experiences and stores them in the brain, compares them, and learns from them. The 'I' only directly views the present. When we carry out our daily activities we do not necessarily have remembering of the present, except as a blur of the past and the present that our brain through practice merges with what we had always considered the present. Yet the 'I' can only exist at the present.

So, the present and the 'I' are one. 'I' defines a present in whatever plane of consciousness we are talking about. 'I' is a product of time and a product of the mystery of 'I-ness' as we do not know from direct experience where the 'I' came from, though we will talk more about this in the chapter on Purpose.

'I' is the zero dimension of time. 'I' defines zero. Without zero, time cannot exist. Zero is that point separating the past from

the future. Zero time is 'I.' That is time without dimension. We define the present as zero, and that is supposed to be the door to the eternal. We now have four dimensions of time:

1) *Zero Time* (Universal Standard Time), integral to all time dimensions. Where 'I' resides.

2) *Linear Time* (our Standard Time) on earth.

3) *Abstract Time* on the Plane of Mind, the realm of souls, time of the second dimension, and the eternal aspect being the Universal Soul, the level of plan.

4) *Spiritual Time* on the level of Plan, defining the third dimension of time, with the eternal aspect equivalent to Purpose.

On our linear time, we look forward and backwards to eternity. We open a door to the higher levels of consciousness by describing circles (describing a plane) or spirals of thinking —expanding our thinking by adding a third dimension to the plane. We also access immediate knowledge, intuition, involuntarily when conditions permit. What we are really doing is placing the 'I' so that it can experience another eternity. *The door to the eternal is in reality the door to another eternity, the eternity inherent in the next level of consciousness (the black hole of consciousness, sort of).*

We can take the statement—the door to the eternal is through the present—and restate it as follows: The door to the next relative eternity is through the 'I.' Does that help? The knowledge of what resides behind that door is immediate and connected ultimately to Will, the power we find at the third level of time, the level of Purpose (or Spirit, if you wish) where our ultimate potential 'I' or self can reside.

The power of Will goes beyond what we characterize as the future and the past. Purpose is eternal. When we say, "That is why," referring to any process, then the *why* ultimately is Spirit, though that connection might seem vague to us at first glance. If we continue to ask Why? to each response, we will end with Spirit (the realm of the speed of light squared).

That refers, ultimately, to That which must reside on the Purpose level of time. That ultimate 'I' or Ego, at the basis of all our experience, forms the nidus around which all experience bases itself; and the quality of Will determines that experience.

When the Yogi says, "I am That," then we know what *That* is. When the Yogi says, "I am That I am," we know the Yogi knows the That, the Ego, determines everything that he/she *is* at the physical "I am" level. The Yogi is saying that he/she sees, and feels, and understands that all experience bases itself on our Ultimate Spark of Divinity found on the Purpose level of time. We are all, ultimately, Sons of God (God being the word describing that which is ultimate or infinite), and all experience express this. Such is the subject of life. And you don't have to believe in God to understand or accept what I have just said. Either way works. Such is the value of the concept of E, in my humble opinion.

I wish I could make all this easier to understand, but I do not have that genius. If what I have just said makes sense to you, then I have done my job. The rest is up to you. You don't even have to remember any of it. If it made sense, you're good to go--even if you don't agree with it.

PERSONALITY

What relation does polarity have with personality? By answering that question, we will take a hard look at the energy components of personality.

Mind exists because of polarity. Thought has much to do with mind, but at its base, thought consists of the mathematical framework of thought. Thought is the recipe; mind chooses the recipes; and experience serves up the ingredients. Mind is nothing but a functioning 'I' or ego, and it surrounds itself during existence with mind stuff. Mind stuff is the thoughts, the thoughts that linger and form the screen through which we view our surroundings.

When we are very young, our experiences cement into the foundation of the personality, and such experiences affect our viewpoint from then on. We deal with these experiences with the equipment that we possess genetically. We all have different abilities and different potentials. As we become adults, personality consists of the mask that we wear, having constructed it over years and years of viewing our surroundings in much the same way. If we want to grow and cast away that mask of prejudgments of thoughtforms, we must face fear: the factor that causes us to erect barriers and close doors in the first place. Evil is the door that closes.

Does that mean all of personality bases itself on fear? No, we could not say that; but we need to examine fear, and when we do, we need to talk about polarity. First polarity.

Polarity

Thought is like a glowing computer system which when exposed to experience, sets about under the direction of the 'I' to organize, remember, and compare by use of that wonderful physical vehicle called the brain. The brain is mostly an organ for memory. Even language is mostly memory. The development of language consists of learning the symbols for experience, remembering them, and then applying them to the appropriate situations.

In the brain's logical functions, we remember experience and set it in order and come to conclusions based on probability. We can also consider sensation as part of memory. The brain brings away something of what it experiences through the senses and records it. Mental visualization is another item of memory. Only when we begin to stretch visualization into imagination and then focus that imaginative power inward do we begin to drop the memory powers of the brain and begin to see the powers mediated by the mind. We especially see this in the process called creativity. *The mind is only a functioning 'I', a focal point for the powers of consciousness, an ego.*

Now to polarity. Polarity results when the ego begins to collect forms of thought through experience. Since these thoughts hang around and affect our behavior, they must have *substance* to them. The essence of actual thought, which is the formula to make thought, we can characterize as elegant circuitry that also is substance. Experience cues the mechanics of thought, mechanics limited by whatever level of consciousness we might involve ourselves; and the thoughts fit perfectly into the brain as the organ of thought.

The 'I' or ego does the thinking and makes the decision to think. The ego makes the decisions. We might say the thinking processes become available to us from the universe, as the universe owes its existence to the processes of the energy of thought. When we have a human body for experience and a brain for storage, then we are ready to go. The ego has the responsibility. When the ego goofs, it pays through its physical vehicle.

Energy, or energy-in-itself, is not something familiar to us. We measure energy or the potential for energy in relation to substance or matter. We do not know what energy is when not related to matter, mostly because we need instruments to measure it; and once we introduce measurement, the energy reacts to the instruments in order to give us a reading.

When *we* energize something, we will always see the phenomenon of polarity. We energize when we create thoughts. We edit experience by feeding aspects of experience into the mechanics of thought and then analyze just enough so that the experience will agree with our preconditions or prejudgments, called thoughtforms. Such preconditions form the basis of our personalities. Then we store all that material in our brains. The stored material serves as evidence for future prejudgments. Sounds cynical? Well, think about it. Eastern philosophy calls the process illusion, maya, or glamour, depending on the level of consciousness.

Let us look at another troubling fact. We cannot face our everyday lives without these preconditions or prejudices. Most adults cannot face any moment in their lives without these preconditions. That's good in some respects. Otherwise we would be overstimulated and disorganized, at least disorganized in respect to the organization of the rest of the world. People would call us crazy. To face reality we must prepare properly, or else face the consequences.

When we energize experience in the form of thought we polarize that experience. We turn that experience into energized substance, and such has polarity. Ultimately, polarity gives us what we call right and wrong, good and bad, the opposites involved in whatever process we might look at. These designations of good and bad or right and wrong polarize all our thoughts that we have used, with good intention, in an attempt to organize our daily lives.

A good person is one with good intention who has organized her/his life to conform with what the culture has termed the good. Many times, we become confused about the so-called good, and we find that the good changes as the culture changes. Then we start asking ourselves the question, what is good? Then we look to religion for direction.

Are religions the storage places for the good? Yes, in most cases they are, and unsettlingly so, because sometimes they make the same mistakes that culture does since religion tends to conform to culture. More reasonably, we would like to see religion as the storage place of worth. Deep within each religion we do find aspects of worth, and these aspects tend to be the same in all religions.

We might have the ability to define some qualities within religion beyond good. The human originators of these religions set forth these precepts to their disciples. We can ask ourselves, are these high rules of religion merely good or perhaps very good, or are they something else? Must we expect the highest precepts in religion to be merely polarized thoughts just like any other thoughts?

We know that polarized thoughts have their imperfections. Not to say that people who have followed in the footsteps of the starters of these religions have done any better. Polarized thought sits at the heart of all our problems. Polarity means difference. The difference causes the hatreds. We work out our differences usually by war. Such is history.

As humans we seemed trapped in polarity. We look to religion to extricate us, and most of the time we fall in the same hole along with religion. Some, very few, do not. It is the exception to the rule that we wish to examine.

Let us say, for instance, that our motivation for living is the aspiration to live the highest ideals as stated in religion. By making that statement we would therefore aspire to the least polarization of thought possible. We must realize, however, that we may not have the ability to resolve polarity within our infinity or level of time (please see chapter on Time if this terminology is unfamiliar). We might have to access another level of consciousness in order to resolve the problem, but the next level of consciousness has its problems, too, except in its infinite projection.

If we want to work on our natural tendency to polarize thought and prejudge most every experience, then we must deal with personality. We wear the mask of personality as part of the polarized world, the world of good and bad. When we begin to talk of polarities, then we find ourselves using adjectives that describe personality. We will now examine personality and how it relates to polarity.

Personality Spectra

Have you ever thought that personality qualities seemed to exist on a spectrum? Let me give you an example. Let us first try to agree on a human value that we would term good in religion and good in any culture and so, maybe, a value beyond polarity.

Shall we try courage?

Courage is the necessary motivation to do what is right, regardless of the resistance. Let us place courage into a spectrum of qualities and see what turns out. We will begin at the end of the spectrum that begins with cowardice. From cowardice, let us start to get more courageous. What is next? Let us put them all in a line.

cowardice<>timidity<>irresolute<>unclear<>undecided.

At this point, we are beginning to notice something, especially if we look up some of these words in the thesaurus. We are now dealing with the un-words, the words that lack some quality. If we continue to try to build up from cowardice all the way to courage, then somewhere we must go from the negative type un-word to something positive. That marks an *interlude*. With that in mind, let us continue further.

From undecided, we will go to decided. Now we enter another realm, the positive end of the spectrum:

decided<>determined<>resolute<>bold<>brave<>captivated<>fixated<>zealous<>fanatical.

Fanatical forms the extreme of the dark side of courage.

Wait a minute! I forgot courage. Where does courage come in on this spectrum? Maybe after brave and before we proceed too far into the dark side. Do bold and brave build to courage? No, not really. Courage is special. Inherent in courage we see belief in something of worth, not something that just might involve tribal loyalty. You can be brave and still do something foolish. With courage, you usually do not go with the group. You are alone. Where does courage come in?

Let's go back to undecided==decided. Something important happened at this interlude. We made a transition from negative to positive, involving a willful act. The ego had to come to the present enough to make that decision. Courage begins there.

Courage begins with a decision. In the process of doing so, the 'I' or ego finds itself at ground zero—between negative and positive. This is zero time (Once again, please see the chapter on Time if this terminology is unfamiliar.). By positioning ourselves at zero, we set the conditions so we can tap the second-dimension of time. An expansion of consciousness occurs because of this decision.

Courage identifies itself as a second-dimension of time quality. That is why religions tout courage. The qualities of the second-dimension of time, courage only being one of them, are the ideal stuff of religion.

We find courage when our will acts, and the ego makes the first courageous decision. We do not find courage anywhere else on this spectrum. We can shake up all of those qualities and extend them in any way we want, and none of them will end up at courage. Courage fits between.

There are many more qualities of the second-dimension of time. We will try another spectrum, that of *compassion*. At one end of the compassion spectrum we will begin with egocentric.

egocentric<>selfish<>unmoved

Again, we must proceed to the less egocentric and we find ourselves at one of those un-words and then transition into the positive. From unmoved we go to

moved<>sympathetic<>liberal<>indulgent.

With indulgent we arrive at the dark side of compassion. Where is compassion in this spectrum? You might say that it fits in after sympathetic. Compassion is becoming one with a person in that person's suffering. Sympathy, on the other hand, is offering pity, a feeling of consolation to the person suffering. Sympathy does not go to the depths of identification necessary for compassion. Compassion is not a necessary extension of sympathetic.

If we go back to the transition from moved==unmoved, then we begin to see compassion. We have an act of will. We allow ourselves to be moved. We begin to feel. We are at ground zero, and the decision to turn around from unmoved to moved not only takes courage, but it is an act of love. Love comes through when we make that decision. We begin to tap the second dimension of time. That is just the beginning of compassion.

The really "good" in life, no matter how good it is, will not plug into the second-dimension or Soul qualities. Good only keeps us on the spectrum, sliding back and forth. It does tend to keep us out of trouble while we are waiting to die, and it does order society somewhat. Good is closer to the second-dimension quality at the zero area than is the bad; but we can certainly see how someone trying to be very good will most likely slip into hypocrisy and end up being very off center.

We see what people might refer to when they speak of the middle road, or the razor's edge, or the position between the dualities. It is that position where we can begin to access the next eternity, the second-dimension of time and at the same time allow those second-dimension qualities to manifest in our own lives.

How exactly can we position ourselves during the course of everyday life? Let me give an example from my own life.

My wife was making cupcakes for an after-school function one afternoon when three of the kids, having just gotten off the school bus, bombed into the kitchen, hungry as bears and just as irritable. They started badgering my wife for the cupcakes, but she didn't want to give them up, first because the kids could have the cupcakes later that night at the school function, and second because supper was coming soon.

To add fuel to the fire, one of our sons had the feeling, at that moment, that he should be top dog in the house, and he started ordering his mother around, demanding things. My wife was frustrated and upset, and I walked into the kitchen midst all this. I had just finished some writing (actually, this chapter).

The tension in the kitchen immediately hit me, and I mistakenly succumbed to it. I said to my wife, "Why do you always volunteer for these things, especially when they interfere so much?"

THE RELIGION OF PHYSICS

Try throwing a little gasoline onto the fire! My wife became more upset, and countered with, "Well, maybe you don't think that your writing interferes!" At this point, we both realized what was going on. We then turned to the kids and controlled them. They backed off.

Let us analyze this situation a bit by looking at an abbreviated spectrum of the word "giving," which was where I was identifying my wife's supposed problem.

stingy<>not giving==giving<>victimization

At that moment, I was stranded at "not giving." (I was hungry, too.) I assumed that my wife was at either "giving" or "victimization." I accused her of being a victim: doing so much for others that she cannot do what she needs to do at home. She becomes angry because she feels misunderstood. I am angry because she disagrees with my evaluation of the situation.

Between not giving==giving we see the beginnings of selflessness. Selflessness is a manifestation of love. It is giving without the expectation of a return. It is *not* giving to end in self destruction, as in victimization. To decide to give, when before there was no giving, that is the beginning of selflessness, a second-dimension of time quality. It also involves love, as do all second-dimension qualities.

I do not know if my wife was being selfless by making cupcakes or not. That is, I don't know if she was making them so that someone at school would say, "With all the things you have to do, you still find time to make cupcakes. Isn't that wonderful!" I doubt that she did it for that reason, but only she knows deep inside herself. Suffice to say, I did accuse her of something that she would not accept. That is missed communication.

In her immediate and angry response to me, she mentioned my writing. She was picking on something that I loved, because I was picking on something that she loved. Her emotional response, immediate as it was, placed both activities in the same category. If she is correct, then her school volunteer activities add up to the same motivation as my writing.

Had I said the following, "You must love doing what you are doing a lot to put up with all that abuse." Then she might have said, "Yes, I do, but these kids are driving me crazy. Can you help?" Such direct communication would have prevented the confusion of feelings and wasted emotional energy. My being more aware would have helped a lot. She says I am like that when I'm hungry. No excuse. In married life, communication cannot be perfect all the time. We learn and move on.

Personality can change with moods and emotions. The emotional entanglements of life tend to mold personality into the prison of thoughtforms in which most personalities reside. If we want to control personality better than we are doing now, then we must know more about personality. Though personality characteristics may reside on a spectrum, the zero point in the spectra represents the point where second-dimension of time characteristics manifest. There we find our higher qualities, and these qualities tend to define what we generally refer to as character, at least those positive aspects. To give you a better idea how Personality Spectrum works, I list below a variety of personality traits on their spectra, with the decision or zero point in the middle (=0=), and the second-dimension quality beneath.

SOME ELEMENTS OF PERSONALITY

Hateful<>Resentful<>Dislike=0=
Like<>Cherish<>Infatuated

LOVING

Cowardly<>Timidity<>Undecided=0=Decided<>Brave<>Fanatical

COURAGEOUS

Depressed-Ambivalent-Unhappy=0=Happy-Ecstatic-Manic

JOYFUL

Egocentric-Selfish-Unmoved=0=Moved-Sympathetic-Indulgent

COMPASSIONATE

FEAR   DENIAL OF FEAR

All personality characteristics have an element of fear or the denial of fear; and that element of fear determines where on the spectra each personality characteristic resides. The associated second-dimension (Soul) quality remains the only personality characteristic *not* touched by fear. The mask of the personality is then removed.

All personality traits are either second-dimension qualities or else perversions of them caused in some big or little way by fear. By way of everyday-experience we pervert the basic energy coming to us from the second-dimension, and we tend to live on spectra of personality qualities.

Fear and the denial of fear come forward as the negative poles of our experience on earth, our experience as animals, basically. Fear serves a positive influence for survival in animals, and we might call fear an involutional characteristic as opposed to the evolutionary characteristics of the second-dimension.

We would all be changing right now in the direction of evolution if it were not for one factor: anger. Anger produces the force necessary to keep a person stalled on the spectra at any particular place. If Divine (Eternal) Will is the force or energy of spiritual evolution, and we participate in that Will when we make the right choices, then anger is the negative force or counter-will that keeps us where we are. Anger includes all the lesser angers such as irritated, impatient, touchy, grumpy, indignant, annoyed, and upset. We see these angers every day, and when we see them, we know we can learn something, if only we would open our eyes to it.

The important point to be made is that if we want to increase our level of consciousness and become more "evolved," we will have little luck in trying to "improve" our personalities by suppressing certain characteristics that we wish to delete. Suppression only results in resistance. This is a law of nature.

We only change by substitution. We need to know how to move on the spectrum to where we reach the ground zero decision, and hold there. I gave an illustration earlier when I described the missed communication between my wife and myself. We all must face the problem of translating the middle road to everyday life. Let us look into that further.

Much of what determines our personality bases itself on emotion, and in many cases, the element of fear. We feel emotions, and our consequent behavior secondary to those emotions determines the tone and face of our personalities. The second-dimension characteristics reside beyond emotion.

Even the characteristic of love (listed above as a soul quality) is beyond what we normally term emotion. Love, in itself, remains the overriding characteristic of soul and of the second-dimension of time. Love expresses the bringing together power of life. We can see all the second-dimension qualities as extensions of love, and when we express those qualities, we express love at the same time. The second-dimension characteristics mediate through the love energy of the soul.

The love energy of the soul is not what nearly everyone terms as love. Most people equate love to romantic love. We do not fall in love, we fall into infatuation. Love, on the other hand, is a power. It grows. It begins with friendship and common sharing. From higher levels of realization love consists of that feeling of commonality or identification we have with all our fellow humans that we are One (a high level of consciousness).

Practically speaking, we can view the duality of personality characteristics to reside between fear and love. Each characteristic has some fear and some love. The esotericist will say that the basic energy is love, but we pervert that love with our fear. All feeling is love, or a perversion of love by fear. Again, the power behind all emotion is love, or its perversion, fear.

Rather than pure love seen in everyday experience, we see the second-dimension characteristic of love as energy but tempered by the mind through experience, resulting in such qualities as wisdom, kindness, joy, and courage. By use of the mind we can apply these characteristics to our lives. As agents for our souls, we serve as the connections between the spiritual and the physical. We become the agents for spiritual evolution. Too much Spirit, at this stage in our evolution, is as destructive as too much earth. Fear attaches us to earth.

Addiction

Personality addicts. Upon the spectra of any personality characteristic we find pleasure and pain. The soul characteristics reside beyond pleasure and pain. Addiction and attachment to our own personality characteristics keep us on spectra. Pleasure, fear, and anger, cement us there. In chemical addiction, especially with the more powerful drugs, the pain and the fear of pain, and the pleasure, keep the person there.

Addicts feel that what they go through is as real as personality, and that is why they can mistake their life for real life. What they feel is real, being a perfect paradigm for what life actually is, at least on the spectra of personality: pleasure and pain. An addict feels that he/she participates in this pain and pleasure most intensely of all people. Such makes them a hero in their own minds. We can also understand why treatment programs that draw on higher values beyond personality, such as AA, have at least some success.

Positioning ourselves at the crucial centerline of the zero dimension of time allows us to transcend personality. How might we position ourselves in everyday life to accomplish our individual goals of self-improvement? We have, actually, no more or no less addiction to our own personality traits compared to the chemically addicted person. We would blush when the addict laughs at us.

Crossing of lines of linear time between two people during an encounter serves as the ideal time to begin a process of examination of personality. Time spent in meditation also has value. I have discussed the types of thinking involved in meditation, but of most immediate importance are the encounters that we have with other people. From such encounters arise the opportunities for service on our part that result in bringing love into manifestation.

The chapter on Light discusses how we must apply our will during encounters, and especially at the very beginning of an encounter with another person to position ourselves such that we become the observers or markers of material produced in the conversation. If something comes up that elicits anger or some other emotion in us, we mentally mark that. We know that all the emotions and personality characteristics within us have elements of fear and love. Anything off the center line, the middle road (the position between the dualities) involves fear in some degree.

Once we mark the emotions, then we know that to issue forth second-dimension or soul qualities, we have to move ourselves on the spectrum immediately to the decision point at ground zero. Such involves an act of conscious will power. We cannot take the time in a conversation to analyze what soul characteristic we are talking about. Nor can we examine whichever personality trait causes the trouble, or how we must move back (or forth) on the spectrum to the neutral area. We have to immediately know how to go to ground zero —that is where the energy exists.

The process of which we speak has a name. The name is *detachment*. The result is impersonality, that is, a separation from the ties of personality. The final result is manifestation of soul qualities. Mostly we use the words detachment and impersonality in the sense of a lack of concern. The meaning here implies detachment from the bonds and limitations of our personality and our well-worn thoughtforms. That takes courage, for instance, revealing something about ourselves or bringing up something with the person that we have been wanting to do for some time, or maybe making an immediate observation about the person which we think is true, but we have not had time to really consider it.

To the esotericist, there exists a physical place that we are calling ground zero. This place locates itself in the region of the head. According to Eastern thought, this area connects with the energy channel that links all the different time dimensions or levels of consciousness within us. When we come to ground zero, we open ourselves to energy inputs from other dimensions.

Does that mean as a conversation begins that we must align our center of consciousness in association with the head? Yes, in a manner of speaking. By so centering our attention, we awaken that area. If we have the ability to hold our consciousness there and still continue the conversation, so much the better. That is difficult. Easier and more practical is to alert the center at the beginning, and then refer back to the center when something emotional is triggered. Such a move allows awareness of our emotion. We do not want emotions to pull us along. We want to stay in charge.

Frequently we will lose the battle. We will fall into passion, which is emotion calling the shots. Whenever emotion has the reins, that is passion. We can look at the emotions as little lives inside us with little centers of consciousness that demand to rule over us. These little lives associate with the thoughtforms that we have built over our lives, and they react with a will of their own. If we fight them, they fight back. We could spend the rest of our lives fighting them and losing—as with the perpetual weight loss program spoken of before—and never go anywhere.

We must go around these ruling emotional ties or thoughtforms. Without the necessary care and feeding (by our stimulating them), they die out. We go around these thoughtforms with the energy from higher dimensions. We access that energy through the ground zero center, consciously disregarding as possibilities any of the emotions of the personality spectra. We hold our consciousness under our control in the center of the head or in the front of the head, according to the effects of our own experiments. In time, we make this alignment subconscious or instinctive, and we no longer need to exercise our will during the process.

Some people, of the more analytical type, might prefer to look at their own personalities and see for themselves what personality traits cause the trouble. For instance, when a person is conversing, and a negative emotion arises, and that is tagged, then the person might want to remember that conversation later in the day and analyze what happened. If we are shutting out a soul quality through some personality problem, then knowing both the problem and also the soul quality involved might aid in heading for ground zero if ever that emotion pops up again.

A few examples:

A housewife has her own way of doing things around the house, and she finds herself becoming angry and defensive if any in the family give suggestions to the contrary.

She asks herself, "What do I think about when I get

angry?" She realizes that any suggestion from anyone about the running the house is met by her own insistence that her way is better. She decides to work-up a spectrum on whatever might be the problem. What to begin with?

She says, Sometimes, they call me bullheaded, but I know I'm not that. That would be the worst way I could possibly be.

She begins with the following: bullheaded—stubborn—intolerant. From there she must go positive: tolerant—flexible—and finally, plastic, that moldable quality that we see in politicians.

Bullheaded<>stubborn<>intolerant=0=tolerant

<>flexible<>plastic

What is the soul or second-dimension quality that fits between intolerant and tolerant? She thinks quite a while about that and decides that to begin to tolerate is like compassion, but somewhat different. To her, the word that most applies is *humane*. To her, to be humane meant to accept the desires of other people, knowing that they, as people, are not perfect; and if we do not accept them, then likely we are not capable of facing up to our own shortcomings.

Now, when that angry feeling pops up, she has decided she should dive to ground zero, the point in the head and realize that she wants to shine forth humanity. She finds that she has less expectations of others, and generally she is happier. She works on becoming more humane for a long time before the quality becomes instinctive.

Another example:

An engineer in quality control becomes fearful when a defect is found in his work. He looks over his work, and he sees that the defect really began in the process before the product reached him. He had not realized the problem at first, and in the last few weeks he compounded the problem himself.

He fears that a superior might call him incompetent. He especially fears that, because he is so independent, so apart from every other professional in the plant. In fact, he prides himself in his separatism. He would tell his friends, "I never associate with those other engineers. They're so square. You know the saying, 'I always wanted to be an engineer and now I *are* one.' That's the way they *are*."

Because he positioned himself so apart, he fears that now he will become a target. He sleeps fitfully for several nights, then he decides to find out what forms the root of his fear. He asks himself what is the quality that might describe the worst part of him in the eyes of others.

He decides that *aloof* fits the bill. He begins his spectrum: aloof—reserved—detached. Then he must turn positive to attached—bothered—overrun. He ends with overrun because he could see so many other people at work who allow work to overrun them, unnecessarily.

Aloof<>reserved<>detached=0=attached<

>bothered<>overrun

What is the soul quality? He decides on *dedicated*. Why dedicated? Yes, he has dedication to quality work, but that is part of the job as he sees it, nothing more. He does not have dedication to engineering, or the company, or anything else that he can think of—except his family. He decides to find aspects of his work that he can dedicate himself to. If after a time he cannot find dedication in his work life, then maybe it's time for a change. That scares him, but he tells himself he will face up to it.

A final example:

A teenager continually says to herself, "I hate myself, I hate myself." Her father overhears her mumbling, and he says firmly, "You don't hate yourself. You only hate the feelings you have."

"So, what!" decries his daughter.

"You are more that your emotions," he replies.

"Then what am I?" she asks. Her father does not answer.

She decides to create a personality spectrum for the quality that seems to bother her. She begins with self-hate, then self-abuse—where now? She thinks for a while and she feels that she is too *negative*. She sees the bad in everything.

She then tries to be *positive* with her friends, but she feels like a hypocrite, and they can detect her lack of sincerity. She realizes that she wants to be somewhere between negative and positive. She sees people who become overly positive (about themselves) as conceited, and she thinks that is just neurotic. She doesn't want that. On the spectrum from positive, she goes to cocky—conceited. The full spectrum:

$$\text{Self-hate}<>\text{self-abuse}<>\text{negative}=0$$
$$=\text{positive}<>\text{cocky}<>\text{conceited}$$

What's the soul quality? She decides on *understanding*. She realizes that her habit of negativity is bringing her down; yet she is dissatisfied with trying to be positive. Neither works. Now, when she begins to get the feeling of self-hate and anger turned inward, she decides not to become positive, but to understand.

She discovered that the practice of yoga helped her. Originally, she entered the discipline of yoga because she thought it would make her more positive. Yoga helped her to deal with the energies, and she eventually learned that her critical eye and powers of observation, rather than being negative, had great value; and she did not want to suppress them in any way.

She became an influential epidemiologist in her adult life, and though she would still admit that she is not an optimist, at least, she is happy with herself (and others). She learned to appreciate the people around her for what they are, rather than for what she believes they should be.

When we can access the energies of the soul and use these energies in our thoughts and everyday activities, then we create a new dimension for thoughtforms. We begin to make forms of thought within the soul attached to the energy of love, not *desire* as in the lower thoughtforms. The soul then begins to develop a body of thought and takes on density. When the soul body becomes sufficiently dense, then it becomes the major source of reference for our daily activities.

We must call these forms of thought of the soul, perfect, at least perfect in reference to our first dimension of time. They represent the times we respond according to soul qualities in our lives. We remember that something has perfection only within its eternal realm. If someone can raise consciousness beyond the soul, then the soul is superseded, not invalidated, but superseded. Any factor located on the second-dimension of time forms part of a more inclusive picture in the third dimension of time.

It appears that perfection has an aspect of relativity, that a quality (and an eternity) within one dimension of time becomes only a part of a greater picture at the next (higher) time dimension. We want to depend on something. Is not eternity eternal, and not relative?

We must remember that the quality of our earth, that is, the tone of the energy of our earth is *love*—at least we name it so; and we know that it reflects the power I am calling Will that created the universe in the first place. We can also see that love has an element of relativity depending on how self-conscious or self-centered that love is. We assume that the Will that created the universe *is* the universal standard of love to which all love is compared.

What can we depend on in our everyday lives? What is

not relative? People who read the holy books of the world will say that we must plug into the Will of the Creator. That is perfectly fine, but what does that mean? How do we plug into that (non-relative) perfection?

We have only one way, and that way is through the decisions (big and small) that we make every day during our lives. We make those decisions at ground zero: the decision whether or not to access a soul quality. Ground zero is the same energy as zero time, the "universal standard time" that connects all dimensions of time. We can depend on that: The Eternal Now.

BEING AND NON-BEING

Being. What is it? And what is non-being? Once a person knows about those things, then what's a person supposed to do?

All words are ideas, and each word represents a separate idea. We cannot find exact synonyms for any word, as each word must stand for a separate idea, or else the word would not exist. Let us not attempt to define *being* by the use of synonyms. We will first examine *being* by saying what being is *not*.

What being is not, is non-being; but non-being is something. Thanks for the clear explanation, I hear you saying. What being is not includes all things and all *predicates*. Our misery and fear reside with things. We fill our lives with things. Things make us comfortable, and so we have comfort and discomfort. We seek something. We want something. All these make up the predicates of our lives. Within the attachment to things we find our misery, and our temporary happiness. *Non-being is that with predicates--predicates being the objects of our sentences.*

Whenever we use a form of the verb *to be*, we say that something is something else. An example: He is a boy. We are attaching attributes to something. He may be a boy if we say so, but we are not drawing attention to the other aspects of his existence. We erase all of the other things and put our spot-

THE RELIGION OF PHYSICS

light on the fact that *He is boy*. We limit when we use verb forms of to be. We qualify and quantify. When we describe anything in life--anything--then we limit.

We prejudge enough so that we think that we are learning more and more about whatever it is that we describe when we use the verb forms of *to be*. In fact, we are conforming our own prejudgments by describing what we can see and understand, when all that which exists *beyond our understanding still remains undescribed, yet present nevertheless.*

Going back to our example, *He is a boy*, what is *he*? *He is he*, of course. Where does that bit of tricky language get us? When we talk of the verb *is*, then we are talking of *being*. There is no possibility that *is* can be correct, totally correct and not limiting, and not prejudicial, other than with an identity: He is he. Everything else subjects itself to error.

Is. The verb lives at the heart of all prejudice and all error, because to use it, begins the error itself. Try having a day without using the verb form *to be*. See what happens. Certainly, if we greatly limit our use of the verb form *to be* in our writing, our writing becomes more responsible. Strunk and White would nod and smile from their graves. I am as guilty as any.

We can take more responsibility for our observations if we only observe and perhaps say, "I see a boy standing there." Then we take the blame for any error. If we say, "He is a boy," then if he turns out *not* to be a boy, then somehow it is *his* fault that he is not a boy. Perhaps he cross-dresses. We could say, "I see a small person dressed as a boy." Heinlein's *Stranger in a Strange Land* comes to mind.

Is. What *is* is? (Maybe I ought to refer to Bill Clinton for this one)

From our observations we cannot use, with total reliance the word *is*. We can fool ourselves, as we may only see part of the picture, and, in fact, we *are* only seeing part of the total picture. Once we launch into the verb forms of *be*, then we subject ourselves to error.

To Be

What about the infinitive form, *to be*? Infinitive? Infinite. Maybe we have found something. To be, as in *The Courage to Be* (Tillich). To be what? We cannot be more than we are. I am I, no doubt. No error there, I hope. We are supported by Popeye's, "I am what I am."

We cannot *be* something. We must be ourselves. That sounds familiar. Then we must use the verb form *to be* without an object, just plane *to be*. What might *to be* mean in that context?

The courage to be: Having sufficient courage to act in a way that demonstrates *be-ness*. No object to the infinitive. *The courage to be* goes as far (up to) qualification and quantification. Therefore, to be, cannot have error in judgement or in sensing. To be does not have the limits of quality and quantity. *To be* has an absoluteness about it. The absolute. Some call God the Absolute. So, *the courage to be* we might say reflects the absoluteness in humans that reflects the Absolute. What does that mean in practical terms?

The courage to be partakes of our absoluteness. Can we have certainty about something that has nothing we can describe? When faced with nothingness, and then stepping beyond cynicism, and doubt, and fear, that does take courage. So how do we act?

To be, or not to be. Death or sleep, as Hamlet pondered. *The courage to be*, as Tillich pondered. *To be* puts us beyond the realms of usual thought, and our misery resides in the *not to be*. **Between the two we find the qualities of the soul**, and we find ourselves faced with the choice of the unknown or the binds of fear.

We have *being*, and we have *non-being*. It appears that both have some sense of exclusivity: two systems. We will take some time to examine systems to see if we can arrive at tools that we can use to analyze being and non-being.

Not to Be

Let us look at any system. In any given system, we find a finite amount of possibilities or reactions. What actually goes on in the system does not define the system. That which lies outside the system defines the system (it's neighbors, so-to-speak). Pause for a moment and think about that.

Any system must have a center of reactivity, also called an 'I,' an energy center of gravity, and from this center we have reactivity in a more or less radial expression to interact with other systems. Examples of centers of reactivity: the 'I' of a person; the purpose of any business (the mission statement); what a machine does (defined by the machines creator(s); what a piece of legislation is supposed to do as per its creator(s). Through the center of reactivity of any system we encounter the axis of the universe (Cite *zero time* in the chapter on Time).

The whole universe? Yes. We must realize that each unit of anything, and each collection of units makes something tangible, in fact everything exists as a unit or as an idea, big or small. Every word we speak, every atom, every proton, every sun, every cell, every person, exists as a *separate idea*, all held together with ideation that includes the entire universe.

Every unit and every system must have a center of reactivity, and that center of all units of reactivity I am calling the axis of rotation of the universe, the same axis upon which we define the purpose of creation, and the same axis upon which we find the energy involved in creativity. All partake of the same energy, and upon that energy the universe revolves. From that (or those) center(s) we can conceive of a system with which we can examine being and non-being.

Postulates Concerning Being and Non-Being

I will state and discuss postulates of the general system of being--not easy stuff--but I will explain the most important near the end of the section.

1. Each idea or unit stands alone but has no significance other than in relation to the surrounding units or systems.

2. What the system is not includes more than what the system is. What the system is not consists of everything other than the system.

3. Two or more units or systems can combine to make a new unit only with the introduction of more energy.

4. The energy used to combine and make a new system goes into the axis of rotation at the center of reactivity of this new form. It requires creative energy to form a new and more inclusive unit.

5. The extense of the sphere of reactivity of a system depends upon what the system is not.

That which a system is *not* consists of everything other than the system: the rest of the universe. That which the system is not contains a *hole* in which the system resides (similar to a black hole? I don't know. Sounds bazaar). The extense of the effects of the system we denote with a circle, called the *ring-pass-not*, which defines the hole in which the system is not. Inside the ring or hole, the system exists. Outside the ring or hole the system does not exist. Everything which the system is not exists beyond the ring.

6. As long as the system has sufficient energy (or fuel), and the forces (in) equal the forces (out), the system will continue to exist.

7. All systems effect other systems in an open fashion. Energy radiates or energy absorbs to affect all surrounding systems in some way. A system affects surrounding systems, and eventually the effect of that system will peter out in space, and we will not have the ability to measure it. I call that the extense of the ring-pass-not. Some systems may only affect those other systems that surround it. Other systems may affect more than its immediate neighbors, perhaps as inclusive as the entire universe. We tend to underestimate the widely dispersed effects of any one system. We realize that we exist a few acquaintances from any person in the world.

8. *That which resides beyond the ring-pass-not of the system defines the system.* That which surrounds the system makes up a more inclusive (a grander) system than just the system itself. We have a grander system surrounding our little system. We cannot say that the grander system negates the smaller system. We would have to say that the larger surrounding system *affirms* the smaller system. The smaller system fills a space of necessity within the larger system.

If the surrounding systems define the system that we are working with, then we have the following corollaries to this postulate (8):

A. *We can define a system by its limits of affirmation.*

B. *A system enlarges or extends the limits of its effects on the strength of its affirmation by surrounding systems--the less affirmation, the greater the growth or extension of the system.*

A universal system requires only the affirmation of a Creator. All lesser systems require affirmation of the surrounding more inclusive systems for existence.

At this point I need to bring up the concept of **entropy**, since it has importance here. **I am defining entropy as the affirmation present in all systems**, and this built-in reactivity of systems (the genetics of reactivity) reflects universal laws of substance--already defined by the ancients: sustaining, destructive, and creative, also called the Trinity. (This is my own definition, and it may fit into yours or not. But I do not know what to call what I am talking about except by calling it something, and entropy fits.). All matter or substance has this genetic makeup. Because of these built-in universal laws, we cannot call any reaction of substance *chaotic*. Even the seeming randomness of the movement of molecules in a colloid, or the course of a stick flowing down rapids, may indeed come under the heading of random movement, but the idea behind the colloid necessitates randomness in order to exist, as does the idea of the rapids necessitates the seemingly chaotic course of a stick floating down a river in order to exists as a rapids.

We are looking not at physical laws here which are *effects*, but at the *causes* to the physical laws. We cannot describe what we see as chaos, because what we *see* is an effect. Entropy affirms, and entropy is the affirmation built into any system.

When we create, we also form a new center of reactivity for this new system. That which resides beyond the creation, what the creation is not, defines the space wherein the creation resides. Again, what the creation is *not* affirms the creation, and that is entropy, at least for this book. I am opening myself to criticism. But, wait and see. It may not make any difference.

All existing substance and matter and all the possibilities of creations have made possible this place in space for a creation, a system. We could say that the rest of the universe and the Mind of God (all possibilities, if you wish) have affirmed our creation. Without this affirmation, without this limitation, we would not have this creation.

We create within the realm of limited possibilities, with those limits defined by the Mind of God or the Thought of the Universe (or the order of the universe): the laws behind the laws of physics. We define God (Energy) in the limiting sense, in saying that the realm of possibilities limits the possibilities. If the intention to create something does not fall within the realm of possibility, then that something will not exist, no matter how hard we try.

Whatever we create cannot have universal application (a perfectly inclusive system) unless what we create has perfection (ideal truth). We seek that. A universal truth has no limitations, and it reflects a universal principle. **A universal truth has no surrounding affirmation--it needs none**. It is. A universal truth extends as far as the universe. Such defines Being. *Being represents a universal truth, relative to nothing.* Such is entropy, also. So, **entropy, the affirmation present in all systems and Being are one**.

9. *Within the realm of relativity, forces oppose one another to limit the extense of the effects of any particular system.*

We associate energy with *Being*, but force with *nonbeing*. Any action will meet opposing forces, and the resultant direction reflects the movement that we call **evolution**. Energy, which we associate with being, *is*--not evolutionary.

10. *The realm of Being connects with each unit at its center of reactivity.*

The process of creation partakes of the energy of being and makes something new, with a new center of reactivity or center of consciousness (same as 'I-ness'). When we seek higher consciousness, we form a new center of reactivity (please see the chapter on Transmutation), and we do this with energy added consciously. When successful, our new center of reactivity defines the center of a wider, more inclusive system.

I realize these postulates have been difficult and, seemingly, disconnected. Sorry about that. Allow me to state the most important carry-away. Energy when applied to matter, which it always *is* to our knowing, is called force, and such is associated with relativity and evolution. If there were energy *not* associated with matter, then that would be called Absolute Being, and that would *not* be associated with relativity. Both are being and both would reside as the axis of the universe. The energy component of each is the Eternal Now. The only problem is that the conception of Absolute Being is beyond our physics, as is entropy. Entropy is a cause, but conforms to the universal laws of substance (what we see), as does Being. I am approaching metaphysics. But I am not there.

Existence

The moment of the present exposes being at each instant, though our realizations of being become fogged with the interferences we place on the present through memory, expectations, fears and feelings of all sorts. These interferences allow us to gloss over the present without realization. A line or curtain exposes the future, and upon that line we all exist, and we assume that we all have the same instantaneous now as the clock ticks along.

Try to encompass the thought that the past and the future all exist as one unit of being or *is-ness*. The future depends on the past as nothing can happen in the future that has not had foundation and roots in the past. Things might happen in the future that would surprise us, but our surprise only exposes our ignorance. With more knowledge of the processes at play, we would have expected that result. Such is the value of retrospect.

We have past, present, and future, but all through this entire process we have *existence*. By living, we are merely going through a process determined 100% by the past, a manifestation of the mathematics of the past. We define the past as an equation that determines the future. We can modify this equation through our actions, but by so doing, we change the equation into another inevitability. Such is the geometry of karma—cause and effect.

Creativity exists on a line of energy present at the center of reactivity of all units, no matter what size, and extends through (the dimensions of) time as the axis of the universe, thereby defining Purpose. Purpose slowly reveals itself through the process we call evolution. Being *is*. Everything else *is not* (yet). To us, the present forms the door to the eternal.

What to Do?

How do I determine what to do next? Does what I do make any difference? Can I see the difference, or must I have the blind faith that it does make a difference? Do I listen to conscience, or do I follow someone's teachings; or do I pursue other inclinations? Maybe I should do nothing--the choice of the catatonic--the pathological fear to make a mistake. The fear of ridicule:

I knew all the time that you couldn't do it.

If you don't have the talent, you shouldn't even try. Why waste your time? You have other responsibilities.

Foolishness. Why not try something very small, and do that really well?

Don't you realize your ignorance? Don't you know that other people have done what you want to do, but they've done it so much better?

I guess I'll crawl into a hole.

Buck up that courage. Surely someone, somewhere, appreciates what you do. Somewhere? Where?

Do we only have a problem in marketing, or do we have a deeper problem: nothing significant to market. We know that illusion markets best.

How about *to be*? How does that market? Marketing has no relation with the *to be*, nor with the courage involved. We cannot market courage. In fact, if we market courage, then it changes to conformity. Courage sets a course, a lonely course, and we see no one around us, except perhaps death on one side and the so-called devil on the other. Nice friends we have. What we *do* makes a great deal of difference in how we view our relationship with the Deity (or the infinite extense of anything).

Transcendence and Immanence

What would be the difference in how we act everyday if we believed in an immanent Deity as opposed to a transcendent one? I need to clarify what I mean when I use the words immanent and transcendent. First, I need to bring back the term Absolute Being. Again, we have interest in these terms only regarding energy. Atheists need not shun away from the word, Deity. I have to use some word to refer to what I am talking about, and you will see what I mean.

You might want to take a deep breath before you go on. It gets sticky (Even more than those postulates?!)

As said before, everything that we can know is *being*, whether we can see it or not. This includes all thought, including the thought behind the creation of the universe. We must agree that the universe is the product of the energy of thought through a responsible organizing entity, perhaps called laws of physics, or call it what you want.

In the conception of Absolute Being there exists something beyond being, the *Being* beyond being, the being upon which our being rests—Absolute Being. That's like the physics beyond physics, metaphysics. Absolute Being is something totally out of our possible conception, but it does have existence, and its existence does not depend on anything, we suppose; whereas *our* being does depend on Absolute Being. Maybe Absolute Being has dependence for *being* on something else, but how could we speculate about that? We should have to change Absolute Being to Nearly Absolute Being.

THE RELIGION OF PHYSICS

The conception of Absolute Being is transcendent, as it transcends all possible human knowledge or conception. It exists in another realm beyond our possible conception. We could say that Absolute Being exists in another dimension of time, but each dimension of time contains all the dimensions below it.

Eastern philosophy refuses to discuss Absolute Being. In fact, it is considered bad form even to refer to it, especially with pronouns, or to conceptualize Absolute Being in any way. We cannot have conception of something of which no conception is possible.

I will define Absolute Being as a totally (to us) independent entity upon which all being depends. We could call this entity King Being, but convention urges us to use Absolute Being. The Easterners would caution us not to call it God. Well, we can choose to call God whatever we want.

As opposed to *Transcendent*, we find an *Immanent* conception of deity much more personal. We can say that there is a God, and that God created the universe. We have God, and we have the universe—two separate things. We might ask, what is the difference? Why cannot this concept also be transcendent? We are not splitting hairs when I say that in the immanent conception of Deity, both the God and the universe have *being*, and the two beings relate, e.g., man created in God's image. This makes God something that we can refer to with a pronoun, traditionally, He (All he's and some she's like that). You can pray to God, and God can intervene in our level of being in a variety of ways. This is religion.

The chapter, E, introduces the conception of God as energy. Using that conception, we can go both ways, either transcendent or immanent. In the *transcendent* conception of God, we could say that God in reference to *our* being is energy, as everything is indeed energy, but that transcendent God is the being behind energy, Energy-in-itself (Absolute Being), energy not associated with matter. Something like the concept of The Word. If this is metaphysics, then we can call God as energy a religion. That's okay. It works. You would have to decide.

E also accepts the *immanent* conception of God, and in that version, we would say that everything in the universe is energy; and so is God. God is energy, too; because God has Being. Anything that has energy has being, and we can identify that energy as force (normal physics), since it has relation to matter.

Do we act differently because of these differing views of reality? That is an important question. I remember a cartoon in the *New Yorker* by Jonick that shows some ordinary dude about to cross a city street. The title was, "The Milky Way (Detail)." That made me laugh.

Going along with this cartoon, let me put myself in the shoes of this man at the curb. I intend to step into the street. Whatever I do, I am heeding my personal will. I *will* some action, and I have responsibility for that action. Before I *will* to do anything, I must have motive. Whatever my motive, I will carry out the action through the discriminative power of my mind with as much love or as little love as my level of consciousness would represent. Whatever I do, even something as mundane as crossing the street, if I am controlling the action, then the parameters of thought, love, and will (the elements of consciousness) must all be present for the action.

If I were accessing a higher dimension of consciousness, then all of my decisions, including crossing the street, would reflect the values of that higher dimension. If I could access the highest dimension, the last turn of the time dimension page, then I would have the power of (a) God. I would be what Christian's term Christ. If that is a possibility, then such is the Christ in all of us, believed by Christianity. Man is then ultimately perfectible, and science can discover all the secrets of the universe.

Just for drill, let us say that I could achieve the ultimate level of consciousness, the last page of time, and I had the power of (a) God. I would still not know *Absolute Being*. Such would exist out of any dimension that I could conceive. I would have to replaced my self-assuredness with *awe*. If the source of all being is *unknowable*, then *being* as we know it and the universe as we know it, must also contain elements that we can never know, unless we can know Absolute Being.

This last statement holds the crux of this chapter. The big jump. Some might say that Absolute Being has the capability of making a perfect system—our entire universe—and we can understand it in its perfection. Yes, that is true, but it is also true that we are dealing with zero to infinity layers of time. Each layer is perfect within its own layer, but only part of the truth when compared to the other layers. The entire system considered all together, with *all* the layers of time, makes up One system.

Is that system perfect unto itself? Or does the presence of Absolute Being, the layer beyond, introduce an element of chaos in the system, chaos, at least to us? The Easterner would say that it does. We cannot know the perfection of the universe unless we know the motive behind the universe, and that motive must reside in the ultimate creator, Absolute Being.

Is there an element of chaos or mystery within the universe that could only be known by Absolute Being? Once again, the Easterner would say, "Yes." The traditional Christian would say, "No." The Christian would say that God created the universe, and both the universe and God have *being*; therefore both are perfect. No room for chaos. Chaos is the word for happenings that can occur but which can never have explanation within any known system. If we could find the system in which these happenings can fit, then we could not call it chaos. According to the Christian, if we find chaos, it is only our ignorance of the system. Who is correct?

If the Easterner is correct, then nobody really can have the keys to the Kingdom. Life has an element of Las Vegas, zero and double zero. When you think that you really know something, then you are only fooling yourself. We cannot be anything but educated fools, ultimately. We must accept a certain element of uncertainty in any system. *Our very existence depends on the unexplainable—actually, the affirmation present in all systems, entropy again.*

Contrast this with the typical Westerner: "Sure, shit happens, but with determination and a positive outlook, things will work out. All it takes is the will to succeed." The Easterner admires that self-assuredness—if only he could have it; but he knows in his bones that he cannot have such confidence. He practices more caution. He wants to see how things play out. He wants a plan.

Here I am, still on the street corner, ready to take a step. What am I to do? If I am the typical Westerner and believe in the *immanent* deity, then I will walk into the street after looking both ways; then I will proceed to where I am going as fast as I can. I need to hurry. Time is money. Hurry, before someone else beats me to it. Hurry so I can go on to something else. Opportunity abounds, and I must seize it.

If I believe in a Deity *transcendent*, then I look both ways before I cross. I also may look up, to be sure something isn't falling from the sky. It *is* possible, you know. Then I cross the street, and I go directly to work. Directly, but I do not hurry. I do not want to waste time, but I will not hurry. If I go fast, I am not hurrying. I must have awareness of what I do, not only to see opportunity, but also to have awareness that I do not make incorrect assumptions. When I am ready to act, then I do so directly, right on target.

The difference between the two we can explain with

two words: faith vs. awe. The Westerner has faith that we can figure out all systems. The Easterner has a natural awe and respect for the system, but he does not have the certainty that we can figure out everything. We must take a broad outlook, says the Easterner.

In the future, these two outlooks, now that the world is shrinking, must merge and synthesize into a new outlook. Regardless how much enthusiasm we might have for our own particular outlook, history tells us that change occurs with synthesis, the merging of two viewpoints into a third one, a new one, reflecting the truths in both the old viewpoints. I realize not everyone agrees with this dialectic view of change (It smacks in Western philosophy of too much Hegel), and if you do not, I can accept that; and I know that you will not accept any of the answers that we now seek.

Synthesis of Awe and Faith

What might the viewpoint of the future entail: The faith of the Westerner that everything has an answer, or the awe of the Easterner facing the great unknowns of the universe? Both will merge. Merge into what?

Let us use the method of personality spectrum as in the chapter on Personality to see if an answer emerges. We will begin with *awe*.

Awe<>wonder<>question<>not know. Do you agree with this? From awe, leading to wonder at the universe, we realize that we have questions. When we have questions, then we must admit that we *do not* know. From *not know* we must go to *know*. Such a transition describes an interlude: Not know=0=know. We then progress to believe<>assume<>have faith. Do you agree with this formulation? From *know* we move away to *believe*, as we admit that we do not know everything, yet. On the basis of our beliefs, we make assumptions in life that reflect faith. We have the following spectrum:

Awe<>wonder<>question<>notknow=0= know<>believe<>assume<>have faith

Still using the personality spectra method, what is the second dimension of time quality that fits in with the transition between not knowing and knowing? What happens in that process? What is the process called when all of a sudden you do not know, and then you know?

First of all, we cannot assume that what we are dealing with is a *second dimension* of time quality. We are already dealing with a second-dimension quality in 'faith'. Perhaps in awe, also. Let us keep that in mind.

I submit to you that the quality of *intuition* fits best. Intuition is sudden knowledge, sudden knowing. It can occur within the Western tradition—remember Newton and his apple? Creative scientists will attest to the value of contemplative thought, equally in the Eastern tradition with meditation and the highest step in meditation called contemplation, leading to enlightenment.

Between awe and faith, we find intuition. To access intuition, we build on our present knowledge up to the point of question. We hold that question in mind, and the answer comes. Ask and it shall be given. Both Eastern and Western tradition accept intuition.

Will such a step make a difference in how I cross the street? So, I am an intuitive. I have direct knowledge. In fact, I might even be an expert in some particular area, but I know that I step forward in my knowledge through intuition. *That begins knowledge.* The details we fill in later. Does that mean that I am standing at the curb, my head in the clouds trying to access higher knowledge? I step out in the street, and I am smashed by a careening Bud Lite truck. Does the next synthesis between East and West produce the absent-minded professor? No. We are talking about integrated personalities accessing intuition *consciously and purposively.* This puts a new twist on intuition, making it something a person can turn on and turn off.

Intuition turns out to originate not on the second dimension of time, but the third. If the formulation has validity, then before intuitive knowledge can serve as the standard of the day, we must first have *integrated personalities* accessing the *second dimension* of time. We can refer to that second dimension as the soul level since the human soul makes up the upper portion of that level (the chapter on Time called it the Plane of Mind).

Personality will still exist, even on the soul level, and in turn, we can use the soul qualities and line them up on a spectrum, just as we have done with faith and awe coming up with a third-dimension quality. Granted these qualities will reside beyond what we might term personality on our level of time, but they will be qualities, just as intuitive would become a quality of the third dimension.

The person represented by this synthesis would be a practical person, but one who relies on intuition. That person would be quite direct when she/he crosses the street, and would certainly have concern about personal safety, but would be alert to opportunities for service. Each contact of the person would have service as its format. The person would naturally serve. The words kindness and compassion come to mind.

If most of the people had integrated personalities, work in life would change. Large organizations would likely still exist. Many people will be happy enough doing certain types of traditional work (as long as they are paid adequately), but the leaders of society, those from which the qualities of the society flow, will not have definable work standards such as we have today. In general, work will have even *more* service orientation, service being the physical manifestation of soul impulses. Work today already is service oriented--if we can't produce something that somebody needs, then we're out of business. Right? Service will take on a new context.

We should have caution not to place present paradigms of work onto something quite far in the future. Much change needs to come about before the leaders of society will be intuitives. Right now, they are politicians, or actors, or both. Some are intuitive. Nearly all are not. Some that are, we destroy and despise. We misunderstand them, misinterpret them, call them impractical, soft. They become the objects of sacrifice. Others run on so-called 'gut-feeling' which has nothing to do with intuition and remains self-centered and dangerous. First, however, all need expertise and experience.

We can begin to see as of this writing (in the 1990's) industries moving toward more service direction, especially involving employee relations, and today in customer relations. All of this moves us in that direction, but the goal sets far to the future, and once we meet that goal, then we will require a new synthesis for a new direction.

The universe is energy. Mass and energy interconvert. *Immanent* deity we can easily see as energy. In fact, a hierarchy of energy levels exists, each termed a dimension of time. Deity *transcendent* we can see as energy, at its essence, the beingness of energy, pure and undefinable without relation to matter. Within E, the striving of humans becomes the science of discovering and using energy, such as intuition, which is the energy already present in the universe. If we view all as energy, then perhaps we can spend less of our energy thinking about the meanings of terms such as transcendent and immanent, and start living.

WORKING WITH A SYMBOL

We are going on a trip with the six-pointed star, the Magen David, the Shield of David, not as the symbol for any particular religion, but as a symbol in itself, Figure (23).

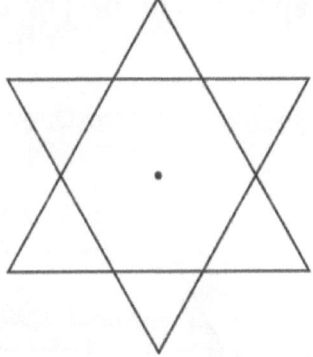

Figure (23) Magen David

The Magen David has been around for a long time, and even before the Christian era it served as a symbol for magic. We have evidence of the symbol in both East and West, even in the New World. Until the 17th and 18th centuries alchemists and religious peoples used it alike. Only in the last few hundred years Judaism moved to use it as a symbol much as Christianity uses the cross. In recent years, the use of the symbol by the Zionist movement has made the symbol solidly associated with the Jewish state. That's nice, but we're doing our own thing here.

Let us see how we might use this symbol in relation to a real situation. A short story follows.

The Pharmacist and His Wife

A pharmacist tells me that his wife blames everything bad that happens to him on his smoking.

"Cause and effect," says his wife.

For me, looking at it, the druggist *is* where he is now because of a chain of cause and effect. That chain appears clear in retrospect, but regarding the future, all the possibilities make prediction difficult.

The druggist, like anyone at any instant, lives atop the apex of a pyramid of causes and effects, his past, Figure (24). He resides at the apex, and his past below him on the expanding triangle.

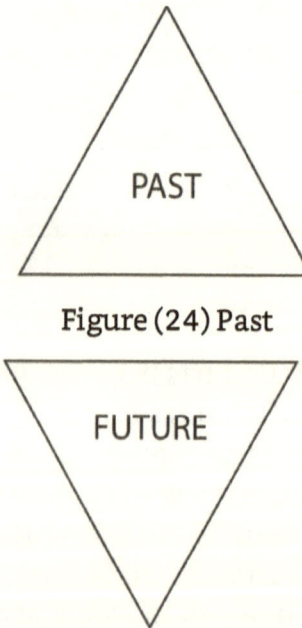

Figure (24) Past

Figure (25) Future

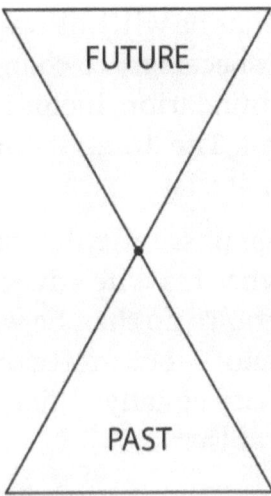

Figure (26) Present

Where he might go from that instant begins at the apex of an additional, though inverted, pyramid of future possibilities, Figure (25).

Within this future pyramid expands his potential as a human. His present, Figure (26) remains an instantaneous point in time and space connecting the tips or apices of the two pyramids, one pyramid above the other, and each connected by its tip.

If he traced chains and branches of cause and effect back to the origins of his smoking habit, he might become lost in the interconnected maze of his character. Possibly he might find the causes to his present effects deep within his pyramid of his past.

"How am I supposed to change that?" he groans.

"Just quit smoking," says his wife.

The druggist nervously taps the tabletop with his fingernail. He has only two choices.

He can begin with the **past** pyramid on which he stands, examine its structure, then attempt a repair job hoping to modify his life so that he will not want to smoke. In that process he might feel it necessary to change much of the structure he built upon his foundation, including his relationship with the people involved. The druggist could decide that his wife must go. He shivers at that.

With his other possibility he could begin at the **present**. He could follow what his wife advocates and refuse to light up his next cigarette. To do that, he would either have to control his desire for another cigarette or else transfer the desire to some other activity equally difficult to control, like eating. He shakes his head at that.

Bored with patterned behaviors that lead nowhere in his life, he decides if he is going to stop smoking that he will *not* take on another addiction.

He chooses to control his desire, and he comes face to face with a self within him that craves cigarettes, his next cigarette in fact. When the strength of this desire-self wanes, he feels good about his decision; but when the desire surges, he must haul hard on the reins of the horses of desire.

He realizes that nicotine stands as his first obstacle. To stop smoking will require control of his craving. He knows that controlling the craving involves only one thing—will. He had heard of people who tried therapy to change their personal approaches to smoking, and that never worked, unless they had the will to stop first. He must apply his will to stop. He shudders, and quietly grinds his teeth. His wife turns her head to see what he is doing. He manages a tight smile.

If he stops smoking, he thinks he will be stronger. Just winning that battle against nicotine will make him stronger. He begins to feel a bit more confident, and maybe, just maybe, after he stops, he will have the resolve to continue not to smoke. What if he still wants to smoke? Will he still have to change himself in some way to stay abstinent?

Faced with the blackness of uncertainty, the druggist still realizes that he is about to make two changes (1) interrupting an addiction and (2) stopping smoking, though he will not see much of the dynamics of his decisions until after he has made them. It is true that applying his will to stop in the face of addiction will strengthen his will power, and it is also true that the decision to stop smoking bases itself on other factors: his health, the concern of his family, the inability to exercise and run with the kids, and others.

Behind it all sits his stubbornness, the independent Marlboro Man still rides: Nobody is going to tell him he can't do what he wants. When he says, No, to smoking, he says, No, to the Marlboro Man inside him. He becomes less oriented to that bit of his desire-self. Anything else he might do after a permanent stop to smoking would be less desire-self oriented, as that self would have been weakened by his exercise of will.

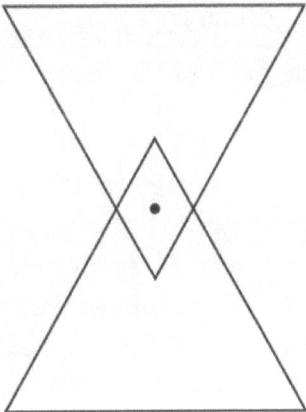

Figure (27) A Decision

Any act toward selflessness begins to open the door to the infinite through the **present**. The druggist's pyramids loom bigger than he might have ever conceived. By deciding less self-orientation, instead of adding to the rubble of the pyramid below him, the druggist has reached into his potential future, toward the infinite, as the future does extend to the infinite (for all of us) even though our life spans remain meager. With that act, Figure (27), his two pyramids begin to move together, apex merging with apex.

His reference as to where he is in time and space remains a point within the figure. Originally a point connecting the tips of the two pyramids, representing an act motivated only by desire (smoking), the druggist's point of self-reference relates differently to the pyramids as they begin to merge, such being the effect of less self-orientation. His conceptual framework or his view of the universe from the point of view of his referent- self evolves to embrace both the past and the potential future. Consciousness rises.

The druggist ponders whether or not he has smoked his last cigarette. His wife stands by, arms folded.

"It would be nice to have clean air in the house," he says.

"Hallelujah," cheers his wife.

He drops the pack of Marlboros in the trash, and sighs.

Life as a Symbol

We would not indulge in schizophrenic thought if we considered the statement that everything that happens, and everything that we see, is symbolic. A symbol represents something universal. That means we can find applications for that symbol at all levels of reality, all levels of being, in all dimensions of time. We sometimes use the words analogy or allegory when we are looking at the symbolic sense of a happening.

Anything that exists, points to something beyond itself. All things relate, like doing carpentry on a boat—work in one place affects many distant places on the boat. Most people, other than poets, have a hard time seeing the larger aspects beyond the physicality in our surroundings.

We tend to call something a symbol if it appears abstract, such as a bunch of lines intersecting in the Magen David. All symbols *are* abstract. The simpler the symbol, the more abstract we consider it, and hence the more general its application.

In the Magen David, we have two equilateral triangles intersecting. Earlier, in the chapter on Plan, I showed how this symbol can represent One differentiating into Three, and how Three differentiates into Seven: a demonstration of one energy changing into the seven forces of the universe. Now we look at the Magen David a bit differently as we try to see how it applies closer to home.

Each and every decision we make every day represents larger issues, the effects of broader and more subtle energies than we generally realize. That is the whole point of this book. What we do reflects what we are, and what we are is a collection of energies. *In each action and each thought we could, if we wanted, distill what we are. As such, each action symbolizes something beyond itself.*

The Magen David, ultimately, symbolizes all our energy inputs. It works at the universal levels, and as I tried to demonstrate (and as I will explain more later) the symbol must also work at the everyday level with each of our decisions. With each decision we make, we face the symbol of Magen David. Let us examine that idea more closely.

In the story, we began with the pharmacist considering a decision from where he positioned himself at that moment. The two triangles touching apex to apex symbolized this problem. As he moved to a decision, the triangles began to merge. Let us see how we could deal with this using some principles brought up in the chapter on Personality.

Once he had conquered his addiction to the nicotine, then he had to face his personality. That concerned him the most. How could he change that? He figured that he started smoking for some reason, and maybe that reason still existed in him. He did not realize that his decision to quit changes the whole system in ways that he could not have predicted beforehand. Allow me to explain this further.

In the chapter on Personality, I explained how personality characteristics reside on a spectrum and that the application of will within the spectrum determines where we find the characteristic that resides in the next higher or more inclusive level of time. We called those characteristics soul characteristics as that was the level next higher that we were dealing with. We have yet to define what soul characteristic involves itself with the pharmacist's decision. So, let us construct a hypothetical personality spectrum to represent what he might face once he overcomes his addiction to nicotine.

The Marlboro Man remained a symbol of some significance. The pharmacist would not admit to that, because that symbol worked at the foundational level, as all symbols must, or else we could not call them symbols. At one time in his life, actually, when he was sixteen, cigarettes represented something that he could decide to do by himself, that he could do regardless of parental dissent, and something that seemed to fit into the peer group that he happened to identify with at the time. Smoking represented independence, as a man, a positive move at that time, certainly a help for him in attempting to resolve adolescence. With that information, we begin to see a spectrum.

When the pharmacist decided to smoke, he was dealing with the issues of independence and dependence. We can identify those two words as the two extremes of his spectrum. In many ways, his refusal to stop smoking for all these years, especially with the badgering of his wife, only brought back those original feelings. Those feelings would flash into his consciousness when he might say to his wife, "I'll goddamn stop smoking when I goddamn want to stop smoking. And not before!"

What was positive for him when he was sixteen, is no longer positive. The positivity has worn off. His fear (of withdrawal) and his anger (not to do what someone else wants him to do) kept him on the same personality spectrum regarding smoking and other areas in his life. We still have him floundering between dependence and independence. Of what else might this spectrum consist?

Would you agree with this formulation: dependence—controlled—not participate. Not participating is the end result of being controlled. When you do not participate then everything that happens is out of your control.

From 'not participate' to 'participate' would bracket the interlude, and it involves the action of will: Not participate=0=participate. To decide to participate means that the person affirms the determination to become a player, a participant. Such a decision involves assumption of responsibility for action. The person no longer can claim to be merely a pawn or an observer.

Beyond participate we might go to disengage then independence. Remember that we deal here with a decision pattern beginning in adolescence, the time when he first decided to begin smoking. We have the following spectrum:

dependence-<>controlled-<>not-participate =0=participate-<>disengage<>independence.

I submit that the second dimension of time characteristic that most fits within the interlude is Cooperation. When he decided to stop smoking, he decided to cooperate. We have evidence of this when he said they should have cleaner air in the house. Such moves him toward less self-orientation, ultimately toward compassion. How might we apply this spectrum to our symbol?

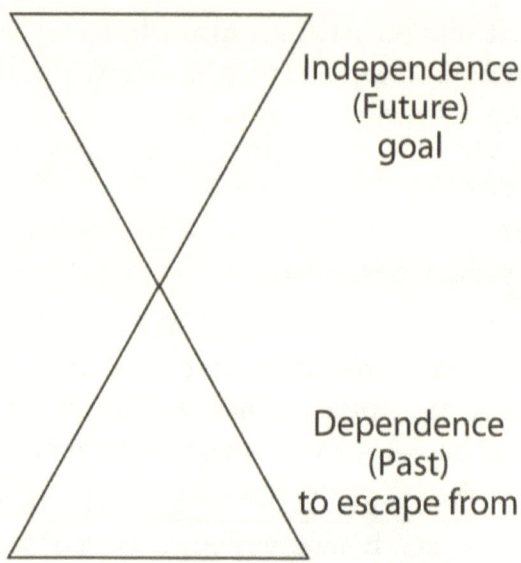

Figure (28) Undecided

Before the pharmacist makes any decision and before he even considers any decision, he resides every day on the extremes of the personality spectrum between independence and dependence: independently astride his Marlboro horse and dependent on his cigarettes.

If we look at the triangles of the symbol as they position themselves before he decides, Figure (28), while he still smokes, then we see the triangles tip to tip. He has made no decision, and the triangles do not interrelate. His life is pure linear time. (Please see essay on Time if this reference is confusing.)

The Independence triangle serves as his Future, what he aspires to, and the Dependence triangle serves as his past, what he wishes to escape from. Neither is a soul quality, but such is the case when we deal with personality. Personality fills itself with false goals, goals that only lead indirectly to the soul, but they do lead there, eventually. When the personality qualities found on a spectrum for any one particular soul quality are mental qualities, then we would call that *illusion*.

THE RELIGION OF PHYSICS

When the qualities are mostly emotional, then we would call that *glamour*. When the qualities are mostly physical, then we might term that *maya*.

Our fluctuations on a spectrum of personality appear like a clock pendulum that continues to swing back and forth, when what we need is a total stop in the middle at the interlude where we find the soul quality. I like that comparison because it takes *will* to stop the pendulum, and stopping the pendulum in the middle would, indeed, symbolize a move to another dimension of time. As we might move closer to a soul quality, we could envision the pendulum traveling less distance, closing in on the middle.

As the pharmacist gets closer to the middle, closer to making the decision to participate, the triangles begin to move together. In such a structure, we have a center, with the triangles beginning to have integration. We do not have total symmetry, yet. Yet symmetry here begins to form. When we create a center of symmetry due to any process of thinking (according to principles spelled out in the chapter on Time), then we can contact, even so briefly, the next dimension of time. In this case, cooperation just begins to appear, Figure (29). On the pharmacist's Personality Spectrum (previously delineated) he finds himself either at *disengaged* or *controlled*.

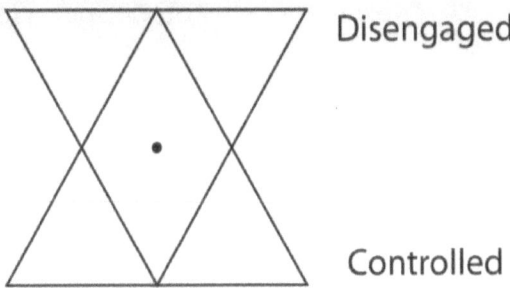

Figure (29) Cooperation Begins

The final stage of participation is the full manifestation of cooperation, Figure (30). In this move, the spectrum integrates and now has a center that allows equidistance to all the possible energies applicable to this particular quality of cooperation. From then on, when he cooperates, he will access some of the energies from the lowest (mental) level of the second dimension of time.

Within that second dimension of time, cooperation will become part of another spectrum that can lead eventually to kindness, then to compassion, and finally to what we could call group consciousness. Group consciousness is 'I-ness' with 'We-ness,' both one. When that level fully integrates, then we can access the lowest of the levels of the third dimension of time which would begin with immediate intuitive knowledge.

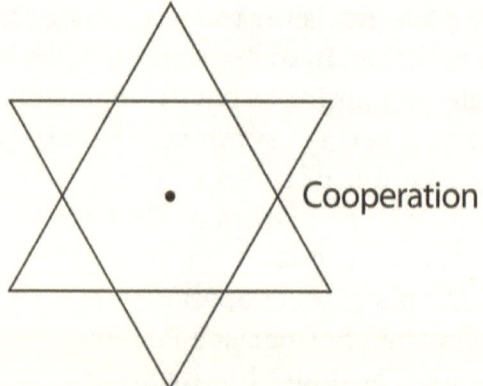

Figure (30) Full Cooperation

THE RELIGION OF PHYSICS

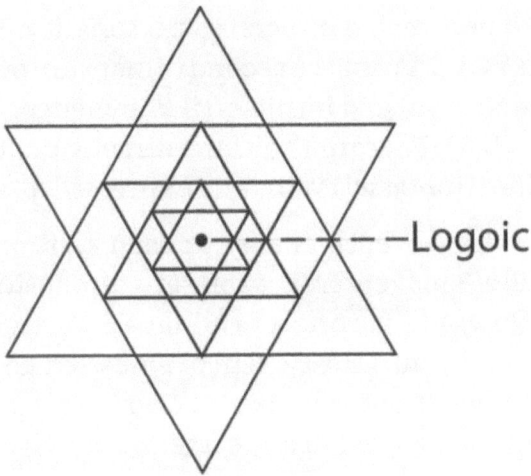

Figure 31 A Human

We can sum up the entire human being with three Magen David symbols, Figure (31), each a different level of time, with the most inclusive containing the lower two, and the second most inclusive containing the lowest. The most inclusive of the levels (Zero time) we could term unity as it represents the only point symmetrical to all.

For students of esoteric studies, the six-pointed stars for each level of time would consist of the following pairs of triangles, with the primary energy manifested shown beneath.

First dimension of time: Astral—*Lower Mental*

LOVE

Second dimension of time: Higher Mental—*Buddhic*

LIGHT

Third dimension of time: Atmic—*Monadic*

POWER

In order to begin to access the second dimension of time, we first must begin to integrate our emotions and thinking to become an integrated personality. The expressed quality is LOVE. From the second dimension of time we must integrate the soul and intuitional knowledge. The expressed quality is LIGHT. From the third dimensional level we integrate Divine (Universal) Will. The expressed quality is POWER.

The seventh energy of each system is the point in the middle. This represents the zero dimension of time, the Eternal Now, and it connects all levels. At the highest level it allows entry into the system of so-called Logoic energy, the energy of what one might term God in this particular system, or the infinite aspect of any system.

If, on the other hand, there are no more Gods (or energies) "higher" or more inclusive than this Collection of energies, then Being ends at the *monadic* level; and what resides beyond, we cannot even speculate. Such territory resides beyond being, as we know it.

The bottom line remains: The symbolic nature of any action we take extends the significance of that action far beyond anything we might consider. Knowing that, we are left with the Courage to Be. Not easy.

TRANSMUTATION

Substance is stuff. If it has any mass, then we are dealing with substance. It is said that substance is *food*. In this chapter I will examine that possibility.

After a person is born, the atoms of the body come from the food eaten and the air breathed. Pretty simple. It tends to become more complicated when we realize the air we breathe is the same air that everyone else breathes, and we humans, along with the other animals, have been breathing and re-breathing this air for a very long time.

Plants put out oxygen and take up carbon dioxide and minerals, and animals eat the plants. So, we interchange atoms with the rest of the world in that way, too. We humans end up being collections of spare parts of just about everything around. We all look different, but we have the identical building blocks as everything else in the entire world.

Do we obtain atoms from any other source? And if we do, where are they?

Let us say that we believe the human consists of more than the body, that because of the factor of the human mind, we have collections of less dense substance *not* packed together as our physical bodies but as substance surrounding us in an energy field. We have bodies, but around the body exists a field of energized subtle substance.

Such is not a far-out idea. Look at the planetary bodies. They have a dense sphere of whatever they are made of, and they also have energy around them; and because of that energy, planets have material floating around them. Earth has atoms of substance surrounding it that originate from itself as a result of the living things on the surface, eruptions from the surface, and particles from outer space that react to it all, especially the gaseous atmosphere.

Just the same, people have atmospheres and reactive substance around them: carbon dioxide and other gases from exhalation, bowel gases, soundwaves from vocalizations, electrical waves from body processes, magnetism, and gravitation (Yes, we have gravity as do all objects). So, let's get serious about the fact that human have auras, to use the term which has been much maligned. The commotion is about what that so-called aura consists of.

How about the energy in the human body? Where does that come from? We all know the energy comes from the food we eat and the air we breathe. I will consider air as food as it does bring us energy, and we do take it in. In our bodies we break this food into simpler components and remanufacture it for various uses. Are these the same atoms that we might find in the energy field surrounding a person?

Atoms of energy? No. We cannot say that. Atoms are little bundles of energy, but they do have mass: substance. Energy interacts with substance to produce force. To us, energy is energized substance, or else we could not know about it. Energy requires substance of some sort to transmit itself through. This is the ancient theory of ether, and I cannot prove that to you, and I will not dwell on it.

Substantial Bodies

We all have various but substantial bodies surrounding our physical bodies associated with emotions, thinking, the soul, and the energy associated with our identity ('I'). When we come into the world, we have all these things. But they are not very substantial. It is said in esoteric literature that we

THE RELIGION OF PHYSICS

are given certain starter units, called permanent atoms of the substances that make up our various energy layers outside the physical body. That just means that whatever is contained in the energized substance surrounding our bodies in human life has beginning seeds of the same substance existing around us when we are born. Again, not a far-out conception. And once the child is born, these atoms attract other atoms of similar ilk. Like attracts like. Still, how would these additional atoms collect over our lives?

When we talk of the nature of these finer substances we talk of mental substance, emotional substance, and soul substance. Emotional substance, for example, would consist of fine substance energized by emotions. The substance part of the substance, other than the so-called permanent atoms, comes from experience. Our experience energizes the substance, but where does the additional substance come from?

Let me reword this question. How can we change or transmute atoms from food into atoms of, say, mental substance? Do we breathe out these atoms? Is it the tone of our words that causes atoms of breath to become entrapped around us and add to the substance already there? Or do we transmute atoms of food into atoms of mental substance merely by thinking? Gurdjieff (through Ouspensky) would say, Yes, to the latter. If Gurdjieff is correct, then we must manufacture the higher or finer (thought) from the coarser and denser (matter). Brain as dense substance produces the electrical impulses representing thought which can energize atoms outside the body to form highly refined mental substance. It would then follow that attention alone can change the character of matter.

Energized substance forms the energy field immediately surrounding the physical body and forms an energy prototype of the body, much as the atmosphere of the earth along with the energy passing through the atmosphere of the earth form the energy prototypes of the biology of the earth. That latter concept is rather abstract, but if you think about it, the energy comes from the sun, and all the biology depends on that energy. The energy from the sun and its interaction with the mineral earth and its atmosphere, prior to the time of any animal existence, determined what life is possible by means of the environment created. As biology developed on the earth, that biology also contributed to the atmosphere and ecology and further limited what type of living things could live on the planet. Let's break out more of those fluorocarbons and see what we can turn into. I hesitate to say, we might all turn into Barbies.

The bottom line is that the atmosphere surrounding the earth is an energy prototype of what goes on atop the earth's surface. An, as yet developed, highly refined computer should be able to examine the earth's atmosphere and then say what is going on with the earth. Same with the energy aura of people.

The human body, like the earth, has its own energy atmosphere, called the etheric body, called etheric as that is the hypothetical media through which energy from the body propagates in order to energize atoms surrounding the body. This energy prototype of the human body contains energized substance and therefore must penetrate the body with channels of energy much as energy from the sun energizes the biology of the earth. The mental, emotional, and the soul bodies are the substantial bodies residing outside the human body.

If we center our attention (concentrate) in a particular area, say at the center of the mental body, then we might well increase the energy of the area, as mental energy or attention can energize substance. Mental energy, after all, is power. Energy will flow, energizing substance wherever attention has focused. We call this changing level of consciousness. Surrounding that center exists a cloud of charged material making up the aura composed of matter of that level of consciousness, for example, mental substance. Before the substance is energized as mental substance, it could be most any kind of atoms. We have zillions of atoms around us at all times. Once energized, it becomes mental substance.

We can move new substance into particular areas with attention. We would move energy in and out of the human body through the vehicle of the etheric body. We also know that new substance can add to the aura from impressions from outside which would consist of energized atoms coming from impressions from other people or from higher levels of consciousness within ourselves. Such is the transmutation process in action. I will use an example with myself as the guinea pig.

A Baby Cries

It's my day off. I've put the baby down for a nap, and I am going to have some free time to read a book that I have been wanting to read for some time.

The baby cries.

My ears send the sound into my brain, and my brain identifies the sound as our baby crying. This is an intellectual function of learning: I have heard the baby cry before. I immediately feel something—how can I describe it?

How about irritation? I want to continue with my great book on lunch menus for mold colonies. She wants me to intersect with her line of time. I guess I can't continue with mine.

She won't stop. I am getting more irritated, but I suddenly stop my slide in that direction, because I know if I go into her room irritated, it only makes things worse—she'll really start screaming.

Walking down the hallway to the baby's room I calm myself. I tell myself the baby is a blessing, and she cries only because she's cold (I had just changed her before I put her down). I hope to find that she thrashed off the covers. I see in my mind a brief flash of when, like a windmill, she casts the covers aside, and how cute she looks. I feel an immediate surge of love.

By the time I enter the room I feel love, and I find her well covered, but crying. I pick her up and hold her close, and I rock her, and she goes back to sleep. I return to my room and my book. But I don't enjoy reading about the molds as I did before, because I allowed myself to become upset, even though I feel okay now. After a while, I take a nap instead.

I will now analyze this little episode in hopes of coming with some insight into transmutation.

The first pick-up of the crying would ruffle my emotional energy field surrounding me, as imperfect as I am. Meanwhile, my brain is stimulated by the sound identified as the baby crying. But the irritation is actually set off before the brain even identifies the sound as crying. When I realize the crying, I am already well on my way to irritation. These things happen in microseconds. My sense of 'I-ness'—my center of consciousness—is awash in irritation. I say to myself, "Don't I ever get a moment to myself?" That only adds to the ruffle.

Through a conscious effort I grab back my center and basically say to myself that I will not allow myself to go in that direction. That takes a conscious effort, an act of will, the most powerful force a person has. It adds energized substance to my center of consciousness. Once I have made that decision, then I feel a flash of love, and I even have a pleasing mental picture of her tousled in her covers. I am able to soothe her and return to my room.

I wasn't able to return to my reading, because my imperfection clouded my reaction at the beginning, that is, my selfishness—an aspect of personality. By over-riding my personality through an act of will, and inserting a positive spin, I added energy to my center of consciousness. I was able to be as a good parent, demonstrate the love that I have. But I wasted energy. I took a quick nap instead and left the molds to another day.

I asked myself why I became so irritated in the first place. What a waste! That is the first step in helping to eliminate irritation from my life and transmuting it into love. That is the struggle we all face to some degree in battling the imperfections of personality, sliding back and forth on the personality spectrum as in Chapter 8.

What Are Emotions

Let us say that I decide that the next time I am faced with something like the interruption of the baby crying, that I resist the temptation to become involved with the ruffled emotional field, and I want to try something different.

"That is NOT me!" I say, "I am higher... Let my wife tend to the baby." Problem solved.

Emotions flow as water. Moisture comes and goes. We know that matter or substance have built-in inertia opposing change, called resistance through friction.

"Okay, okay, I'll go get the baby myself!" I yell to my wife who is busy doing something down cellar.

Emotions have come about for a reason, let us say, a good reason. Emotions serve as an extension of the physical body, as a relatively primitive method of response, quite short of a reasoned response, more of an instinctual response much like in animals, except we have the added power of our intellectual brains. Emotional-type material such as fear and parental attachment have positive evolutionary value in animals.

Emotional responses do not involve thought, but the important thing to remember is that they will attach to thought. In fact, emotions have no way to imprint themselves for any permanence unless they attach to some sort of thought. The ancients characterized emotion as water, because of the inability of emotion to have conceptualization just by itself. Allow me to explain this further.

We conceptualize thought by the use of symbols: We can write or talk. The spoken or written word conceptualizes thought. How can you conceptualize emotion? We cannot, in words, describe fear or any other emotion. Art can do it. All art taps into the emotional arena in some way, but thought is NOT emotion.

Emotions consist of agitation of emotionally energized substance. Life experiences attach emotion to thought. For example, we have an experience that hurts us, and we learn to fear having that experience again, and we watch out for it, waiting for a similar experience to appear. At that point we have created a thoughtform. A conditioned response to that thoughtform will unleash the emotion.

Dreams conceptualize emotion. A dream is existence on the so-called astral or emotional level of a low order, without the 'I' calling the shots. Thoughts are things in dreams, and the tone of the dream tends to portray emotion. Not having an 'I' in control, dreams will portray us lost in whim and impression. But such is a necessary function in our existence as it works out the emotional perplexities of our daily life. Without that, we go nuts. Let me briefly explain that last statement.

If we waken in the early morning hours, we feel concern and many times fear about things we think might happen. What a terrible situation we are in! I call it the 2 am horrors. Then we go back to sleep for a few more hours and dream. That dream-work allows us to come to terms with our fears (we need not remember the dreams); and when we waken, we wonder what the fuss was all about. In turn, people who have sleep disorders or who are taking medications that interfere with dreaming can become increasingly anxious and depressed, developing pain syndromes. That is life without proper dreaming, and in my opinion, such is the origin of much of the anxiety and depression I see in people in my own practice. I have said "We" in the above example, because I believe this can affect any of us. How do you feel about that?

Transmutation of Emotion

When we work to control emotion, we must fight the power of conditioned response—much like Pavlov's dogs. Very difficult. Emotion clings to thought in order to have permanence. Thankfully, through thought we can release ourselves from the tie to negative emotions, but we need the energy from a higher level of thought to energize other more positive areas.

When we have the ability to come to the present and realize what we are doing, that brings us back to 'I' with the ego controlling. The heat or energy of the 'I' exceeds that of the emotional substance in turmoil. There's a sudden flash of love, as in my moldy example. It takes effort and time to substitute thought, higher thought, to decondition the conditioning of disturbing thoughtforms. These thoughtforms keep us from the soul characteristic in the middle (the interlude), so to speak, of any personality spectrum we reside on.

Once freed of the prejudice of a thoughtform we can enter a new dimension or perspective (not necessarily a totally new dimension of time, as there are many little steps within any level of consciousness). Once having gone through the realization through understanding, and we begin to view reality in another perspective, it is possible to feel a bit plagued by the fact that we must have been practicing and bringing the mind to the 'I' and adding energy to the thought content associated to the 'I' for quite a while. We can feel almost lost, maybe that we failed, no longer irritated, but lost. During that period, the polarity of the self establishes itself in a new realm, a new perspective. Facing the blackness of an unknown step, a step into what we do not know, that is NOT the time to lose nerve. This is what courage is all about. Each step of change, whether we are dealing with learning, understanding, or even the big step called conversion, or initiation, all phases of change involve the transmutation process. Transmutation is the blueprint of change. The steps:

1. Fear of change, awash in emotion.

2. Exposure to another way, and frustration that we cannot be that way, knowing something better exists.

3. Energy added to the system in form of insight. Gradual breakdown of the old system ensues.

4. Uncertainty of the unknown in the future. Dread. Thoughts about turning around and going back. (I know I had that during my first winter in Maine.)

5. Leap of faith into the unknown, transitioning into the new perspective with all the anxiety that might attend that leap.

6. Realization.

7. Reorientation to the new perspective.

8. Attempts to convince others of the validity of the process. This can be quite destructive. Best to keep the mouth zipped.

Inner growth, in any respect, becomes nothing more than spiritual risk-taking. We can look at it from the outside or the inside. From the outside it is modern psychology. From the inside, it is called esoteric psychology and ever so much more complicated than my minuscule examples.

At each major level of consciousness that I define as

THE RELIGION OF PHYSICS

another time dimension (if the change is all encompassing), Spirit expresses itself. Spirit is energy not attached to matter, and Spirit remains the energy behind all the elements of consciousness—love, thought, and will. The horsepower (Being applied to matter) of Spirit is love, hence the realization that God is energy, which is the realization that set this entire book off in the first place.

Change, at whatever level of consciousness, can bowl us over at the beginning. We can feel awash, at sea. Takes time to define a new center of consciousness (an 'I'), increase its energy and discriminate that center from the sea of other incoming forces.

Once we can discriminate the self, then we approach the next level onward. At that point we access help, and energy appears in a positive way; and we learn to recognize it. We get help? Wait a minute, you say: You mean that someone gives us help? In a way of speaking, there *is* light on the path. There's an important question from the bleachers: Is someone or something holding that light? The rest of this book will be about that. Hang on. And I haven't forgotten you materialists. There is room for all.

Whether there is help or not, we begin to break up the attachments of thought to emotion, saying goodbye to that stuff, loosening both the center and the surrounding substance to seek something higher. All during that process, the energy of Spirit becomes more apparent since we are viewing it not so distracted by matter. The transmutation process enlightens us to finer energized matter. So fine, in fact, that at some time, we lose all individual identity and merge with the universal force. This is called a tantrum—best exemplified by a three-year-old. So much for evolution.

The transmutation process is a paradigm for change within any system. It is all about energy and matter. Substance, indeed, is food, and its origin is food. Humankind has gained the role of transmuting or digesting substance and energizing it in relation to the entire biology of the earth. Humans serve as an important part of the biology of the earth, but only a part. Kind of a big responsibility. We remain part of the biology of earth.

Psychism

Let us look at a subject which fits in-parallel with transmutation, that of psychism. Psychism is a study of sensation outside of physical sensation.

When I earlier mentioned the aura of emotional, mental, or even the soul substance that surrounds a person, I was, obviously, referring to energy fields of substance. When information transmits into an area of an individual, it consists of any of these kinds of energized substance, and the corresponding areas of the aura, or energy field, will have stimulation. Like stimulates like. Not always will we have consciousness of this stimulation—not that it is an unconscious transmission of energy, it is only that we do not have the realization of it—usually being distracted by everything else that is going on at the same time. Nearly everything we term as unconscious is better characterized as *briefly conscious* since we do not realize it nor catch it. I do not believe much of the unconscious.

We brush close to these stimulations coming through and disturbing the aura of energy surrounding us. We might say, "I felt the tension in the room." Or, "I knew that was going to happen even though they didn't tell me." Occasionally at higher levels of realization: "I finally understand: a ball is a sphere!" Big things like that. But I think all of this is obvious enough for you to share such experiences from your own life.

Psychism is the awareness of information coming in from sources outside ourselves, not mediated by the five senses that we all know. Acquisition of psychic powers goes along with the transmutation process, as during that process we break apart thoughtforms, and we consequently realize the subtler psychic transmissions, faint inklings at first.

So-called primitive peoples have such inklings more strongly felt as they have not had generations of mental interruption and complication that interfere with transmission and reception of subtle energies. This mostly relates to the extension of the emotional senses, the same in animals, but it is said that it also includes thought transmission. I believe that is true, but I cannot prove it to you. What do you think?

An animal senses emotion by instinct. Thought also exists in animals, as they can learn the same as humans when exposed to harmful or helpful situations. Over long periods of time this learning eventually turns into instinct. For example, learning can begin with a harmful experience with consequent fear of being hurt again, recalling that experience through the formation of a thoughtform. The thoughtform builds into the species through natural selection since its aids survival. Over eons, birds automatically fly away from foxes, and fish swim in schools.

We humans also have instinctual thoughtforms left over from our primitive pasts. Most thoughtforms we make ourselves, and others the culture instills in us as we grow up. Cultural thoughtforms blind us, though we never think about them since we have always lived that way. Traveling to other cultures can get us to thinking about our own thoughtforms. Travel enlightens.

Concerning psychic powers, we lose the lower ones the ones animals have as we civilize ourselves. That does not mean that we do not have these powers potentially. We CAN access them, I am told. Cite so-called black magic, which I know little or nothing about. Am I being ignorant in saying that I cannot see anything useful in a positive sense in trying to develop the lower psychic powers?

Concerning higher psychic powers, we will have inklings, tendencies, and occasionally intuitions. When applied to everyday life, those powers express as love and benefit humankind. At origin, psychic powers base themselves in the energy of Spirit, though in the lower powers they mostly relate to self-survival, cunning.

Cultural Thoughtforms and Evolution

I have discussed transmutation in relation to the individual and then in relation to psychic powers. We saw how thoughtforms become semi-permanent in us and how they cloud our ability to understand. I mentioned briefly the clouds caused by culturally induced thoughtforms, and now I want to explore that more. We can break up culturally induced thoughtforms as we would any thoughtform.

I suppose I could use the word preconception instead of thoughtform, but thoughtform emphasizes the link to emotion, whereas preconception relates to 'concept,' which is more of a mental thing. As Mark Twin said, "You can't reason someone out of something they weren't reasoned into." I take that as referring to thoughtform, which has staying power since it is tied to emotion, and it requires, not argument, logic or reason to resolve, but substitution of a higher understanding within a higher level of consciousness, one *closer* to 'We all are one' rather than just logic. Concept remains a more mental thing, subject to logic. Perhaps I am splitting hairs.

The subject of evolution of powers and substance brings us to the question whether or not we can see different *levels* of evolution within the human family today. We can see from our own observations that some people have traveled very far down the path of spiritual evolution. Can a country, or a race, or a culture claim that its evolution makes it more advanced spiritually than others? Has the transmutation process worked equally or unequally among different peoples of the world?

We could plop ourselves into the middle of any foreign culture or domestic subculture that we might name, and if we had to survive as they do with *their* tools, then the people of that culture would quickly see us as deficient. (That's how I felt when I first moved to Maine from California in 1971.). Cultures define the normal through the tasks of daily life. If we transported ourselves to the middle of the Brazilian jungle, our knowledge of the surroundings and the skills necessary for survival would make us look silly. The natives would laugh at us and likely help us out of pity.

We acquire knowledge through learning. That is true in all cultures. The human being, by the nature of the species, begins to learn despite the environment, whether the environment is an American living room or the Gobi Desert. Depending on the harshness of the environment, much of that learning goes into survival. When daily survival looms not so difficult, then the learning goes into other areas—like who has the highest batting average—things like that.

The point I am trying to make is that the capacity of the brain to learn is equal in all humans, but culture, conditioned by the environment, determines the content of the material learned. Anatomically, the brains of any particular racial grouping are not different. There is no species difference within the human race, though when some people use the terms black, white, or brown they tend to conclude a species difference. We have only one race: humankind. I have to admit that there do exist genetic differences in various races concerning susceptibility to certain diseases.

Evolution in humans relates to the development of the mind, but the brains remain equal. We find learning in any culture, the only difference being what is learned. For us to learn how to hunt caribou and use all the caribou parts most efficiently would take a long time, just as it would take an isolated Eskimo much time to think in the abstract mathematical symbols of algebra.

Racial and Cultural Transmutation

Humans have evolved quite quickly, and we face the question whether or not some human cultures, based on racial lines, would have instinctual thoughtforms resulting in (genetic) behaviors and skills particular to that culture. Allow me to give you a mundane example.

Three of our children are from India. I talk with other parents of adopted Indian children, and I have heard the following frequently enough to comment on it: "The Indian girls are great with their hands, so quick and dexterous, so graceful." If such a statemen is true, then the grace and dexterity is not a culturally induced trait, because they are growing up in American culture. Whoever called Americans graceful?

Are we talking about a possible racial trait? If actually it *is* a trait, is it a racial trait based on thousands of years of homogeneous culture, a culture when dexterity of hand served as a valuable resource, one that in hard times would impact survival?

What is the answer to that question? I know it has importance, because some people are waiting in the wings ready to say that certain racial groupings are inferior because of inborn traits, negative traits, just like the animals: fierceness is positive if you are a badger, but negative in most situations if you are a human. They would say that humans are not a homogenous species and that races differ. We speak of cultural diversity, and they speak of racial diversity.

The idea that one's own culture or grouping is in some way superior to all others is called ethnocentrism. We call it racism when applied to race. The ethnocentric outlook is ancient, and it is present in all peoples in its distribution. We could call it part of human instinct, and instinct left over from the animals. Translating ethnocentrism and racism to the animal model, they connect to the herding (like with like) instinct and serve as a positive aspect in animal evolution, but no longer in humans.

We use the ethnocentric instinctual thoughtforms in the only way we can: we distinguish between races and cultures, which is okay in itself, but we go on to say (not always to say, more often, infer) that whatever we have is better than others. We can even use our own tests to prove our point.

I am saying that *all cultures and people must use the transmutation process to learn and grow evolutionarily*. Then we turn around and say that a native of Borneo cannot use the transmutation process for evolutionary ends to the same degree as a native of Orange, California. Sorry, that line of thinking will not work. We already have inferred equality in brain capacity and learning in both environments. Therefore, the transmutation process works equally in both. But to what end? Which is better?

Better? How can we ask such a question? Makes no sense. Whatever trait that may or may not build into any race by transmutation is a positive trait, or else it would not be there. You could ask why is this different from ethnocentrism? Ethnocentrism is a dysfunctional quality in humans, and we call it instinctual, do we not? At least I do. I had said previously that the herding instinct was animal-related and very old. True enough. There *is* genetics in tribalism.

The key is that animal instincts still in humans are, indeed, very old. In fact, they are so old that everyone has them embedded in the genetics, and not more in one group than another. Instinctual traits that would have arisen once people had differentiated into races would all be positive, if indeed there are any, because they would have arisen by the transmutation process. Transmutation, when compared to animal instinct, is a positive turn.

The Soul, that center of energized substance that defines the 'I,' manifests ultimately through the transmutation process. That is the direction of evolution, and it remains beyond the capacity of animals. When the soul manifests in a certain way because of cultural conditions, then we see facets of quality of the human race. We do not see traits appearing that are better or worse than traits in other cultures. We see different facets or qualities of the soul manifesting with the environment having much to say about which traits have importance.

Given equality in the transmutation process going on in all races and all cultures at the same time and given relatively the same beginning time for human evolution, then humans when divided into cultures or races come out to have equal evolution as human beings. The animal in us is the same in all groups, as that was very early. Transmutation is more recent and *always positive* in human terms.

Within individuals within cultures and races, we will see the gamut of abilities and see different levels of evolution. Some people act like animals, others like saints. Some only watch television. We could very well classify individual people into evolutionary groupings, but those groupings would go across both cultural and racial groupings. Such are the "races" talked about in esoteric literature, and they have relation to evolutionary status and not to race as we ordinarily use the term.

Comparing Cultures

Evolution of humans regarding the transmutation process is one thing, political power, industrial and military strength quite another. It is a given that the cultures that have the ability to build the greatest war machines think they are superior. To win at war, it is thought, requires the will, the intelligence, and whatever else it takes to make the winner the best (highest) civilization. The type of learning that went into the making of war machines is of a particular kind of intellect and particular type of learning. It is not all learning, but we have placed a premium on that type of learning. We say that our learning is better, and we have learned the right things. Might makes right.

Our problem is testing. Testing is supposed to help guide learning. Instead, people look at winners and losers in testing. The old herding instinct raises its head again. People use tests to grade people into smarter and dumber. We first have to ask ourselves why we want to test? Then we have to ask ourselves what qualities do we miss when we test?

If the test is a prerequisite for training of some type, say for admission into a certain type of school, then the makers of the test have already formed in their minds the most desirable type of person they want for that type of school. The test should ferret out those desired people from the herd. The school will then only admit ferrets. Actually, the school will then try to admit only those who scored well on the test, and this process would include personal interviews, and the interviews where an 'ideal' type for that training is sought.

We begin to form traditions. The traditional doctor-type. The traditional lawyer-type. The traditional teacher. We establish cultural norms. Then we wonder why we have trouble. The professional organizations perpetuate these norms. Over time we not only see some of the positive effects of such norms, but also the negative ones. Professional schools become the bastions of conservatism. New paradigms are looked upon with fear as they might change the comfortable status quo.

What are some of the things that we might miss when we resort to intellectually testing people? Allow me to list them:

1. Higher and lower psychic powers.

2. Depth of ability to love.

3. Personality integration

4. Ability to access soul qualities

5. Intuitive abilities

All the above qualities reside in a different place than pure intellectualism. In fact, when we emphasize the intellect, many of the things on the list merely bury themselves deeper, away from discovery. I would venture to say that many of the cultures that we consider primitive are the cultures that well express areas on the list. They have been learning those qualities through the transmutation process along with the skills of survival. We see that difference when we realize that the 'wise man' heads the tribe. In our culture the most political and powerful leads the tribe. Maybe with modern times, the two are merging.

We cannot say that just because a less technological or more primitive culture has manifested more of the aspects on the list than perhaps our own, that this culture is therefore better than ours. The culture can manifest certain aspects of Soul, for example, better than our own because of their simpler ways (sort of a monastery concept). When we concentrate our efforts toward our cultural goals, we do not have the necessary attention or sensitivity or time to pay attention to Spirit. We do pay attention to Spirit, but we do so much else that it sorts of gets lost in the mutter.

I can understand people who feel estranged from our culture and its political system, and they see our culture and their own development going in different directions. Strange as it may appear, the new paradigms for our culture come from such people. We need to listen to them, and they need to hang around, rather than flee to, say, Canada. Yet, I can understand that. You gotta be happy.

We ask an incorrect question when we wonder if one race or one culture is superior to another. Instead we should ask a question something like the following: Since culture and race seem to go together as seen in foreign cultures, how can we best learn from them as we work to evolve the particular facets of human potential that we wish to express in our own culture, knowing full well that all of humanity is one. It adds up to the following:

Race itself is but appearance.

Culture is something, but it evolves.

Soul is something and expresses the same in all.

A hard pill to swallow but let us all take our medicine. None of us has perfection. We learn from one another. The soul will out, eventually. Love, the great energy and quality associated with the soul, also powers creativity. And it turns out that the most creative in energy terms is the most efficient in evolutionary terms.

Concepts to Carry Over

An energy field of substance surrounds each of us just as atmosphere surrounds our planet through gravity, a mutually shared quality. The energy activating the substance surrounding us comes from the consequences of our thinking and acting through the elements of consciousness with our own minds and from higher levels of consciousness, and from other people. A story shows how a father transmutes irritating energy generated from a baby acting like a baby into the energy of love.

Emotions are energized emotional substance. All emotion comes from love, or its perversion called fear. Fear worked fine with animals. Emotions become attached to thought as thoughtforms, and we mentally file them away for when we need them.

We transform thoughtforms through the transmutation process. The steps in the transmutation process can be large (called conversion) or small, but they remain the same as in any growth of Spirit or consciousness, or resolution of so-called neurosis. This is the study of esoteric psychology.

Substance has function as food. Energized substance is also our Spiritual innards at all levels of consciousness. Ouspensky says it best.

We have lost much of our primitive species psychic ability because of our civilization and our complicated thoughtforms. We have mostly under-developed 'higher' psychic powers associated with our potential for the higher levels of consciousness, and these powers mostly relate to intuitive thought.

Cultures have embedded and treasured thoughtforms. Brain capacities are equal in all cultures, but cultures determine what should be learned. A culture uses its tests to perpetuate cultural thoughtforms.

We all started out the same evolutionarily in the distant past, and so-called human nature relates to our animal-like pasts. The more recent additions in human consciousness have occurred exclusively through the transmutation process in the which Soul qualities become better expressed. Any culturally related racial traits are therefore positive traits. New paradigms come from people seeing through cultural thoughtforms. The energy of love is the most adaptable evolutionarily.

I am going to further explore psychological mechanisms in the next chapter.

PSYCHOLOGICAL MECHANISMS

When we consciously work with energy and realize that energies interact to produce everything we feel, think, or do, we begin to see psychology in another light. Such an orientation with the focus on energy is called esoteric psychology. Esoteric psychology already is a highly developed field, more complicated than traditional psychology and psychiatry, and more esoteric than most people would want to tolerate.

We would do ourselves and our successors great injustice, however, if we simply make the decision that traditional psychology is now passé, and that we should substitute something else for it entirely. Traditional psychology and psychiatry have much to offer. We can supplement traditional thinking with the application of some esoteric principles.

In this chapter we will begin to apply the science of energy to the concepts used every day in psychology and psychiatry: neurosis (I freely use this 'old' term, because it is useful), personality disorder, character disorder, and schizophrenia. Such only serves as a very first step. Anyone interested in pursuing such studies has a road ahead requiring education that can last a life time.

We have no standard textbooks on esoteric psychology. The only book by that name was written over 80 years ago, at the infancy of modern psychology. What I say in this chapter will agree with some existing works and disagree with others. Hopefully the material will give you, the reader, the tools to begin to understand the field and incite interest to pursue it further on your own. One thing is for sure: If you have the fortitude to actually read this entire book, you will likely develop insight into energy and how it affects our lives, and you will have the ability to look at traditional psychology (and life in general) in relation to energy. Nothing in this book should be taken as 'Scripture.' The goal is perspective, leading to intuition—on your own terms.

Let us first talk about neurosis. The word, neurosis, has been around a long time, and many people use it in everyday conversation and cannot precisely define it. Modern psychiatry defines neurosis in its diagnostic and statistical manual and describes the many different presentations of neurosis, such as depression, anxiety, or compulsions all as separate diagnoses, highly refined, and each having a different number. Most of that has to do with billing and statistics. Nevertheless, neurosis is still here, and it exists. I discuss neurosis only in regard to personality spectrum, which is my own creation, for better or worse. I do think it has much to offer.

Primarily, neurosis involves anxiety, the fear of loss of control, the sublimated fear of death. When we talk about neurosis, we are talking about fear or the denial of fear. We resolve neurosis as we come to grips with what we fear. When we discussed personality integration and personality spectra, it was shown that fear kept us in any particular personality configuration. Fear prevented us from stepping to ground zero, touching the soul quality, and growing. In alignment and integration of personality, faced with an unknown, we make a conscious decision, and we begin to access soul qualities. We move toward ground zero (the interlude), the area where we *can* access soul qualities.

I want to make the specific point that the process of

resolution of neurosis and the process of personality integration are the same, the same mechanics, merely different names. I am going to attempt to show how we can view both processes and how they equate. To do so, I will have to discuss psychodynamics and become a bit detailed. Even as detailed as this material might appear, we are only at the very primary level of looking into this subject. The field is vast.

In neurosis we suffer because life does not flow according to expectations and hopes. We see what we deem as reality, and we find that we cannot control reality as we might wish. If we want respect, we get derision. If we want love, we get stepped on. After a while we learn that the only way we can continue to live and not shoulder hurt and disappointment all the time is to construct an illusion of omnipotence. We satisfy ourselves that we are what we want to be through an illusion. Not such a terrible strategy, as we will see later.

None of us lives in isolation, and our desires, be they wealth, popularity, inner peace, even enlightenment, are nothing but desires. We derive our desires from what we see around us. We see others who may have the qualities that we seek. We want to be like them. When we see the long road ahead of us to get to where these people might reside, we feel discouraged. If we try the road and fail, we start erecting the standard neurotic defenses.

It is like a race, only we are losing the race. We see life as competition. When another person comes along who has the qualities we seek, we reject that person and find fault rather than seek out the person to discover how she/he has succeeded. To seek out that person, we would have to admit to ourselves that we didn't have that quality we want. We figure we must have something the matter with us that we do not have that quality, and the other person does.

Other times we will seek out people who have the qualities that we imagine we want, and we will idolize those people (the flip-side of rejecting them). Media stars serve as an example. When we idolize an actor or an athlete, then we identify with a quality that we seek; and it makes us feel better. We generally add a powerful dose of day dreaming, putting ourselves in the place of the star, and through that mechanism we create an illusion of omnipotence. When the star may be exposed through the media as a sham, then we feel betrayed.

Fear expresses as indecision and dread. The key to neurosis is denial of fear thereby creating a feeling of omnipotence. We see this in adolescents prior to the resolution of adolescence. The adolescent in a developed country does not feel the breath of death in life, ordinarily. The adolescent approaches life as if protected by a charm. Such an attitude *in the adult*, makes for irresponsibility: We feel that we have plenty of time to amend wrongs, plenty of time to allow others to forgive mistreatment. Still feeling much as a child, the adolescent (or the adolescent-adult) refuses to see the ripples and waves that he/she causes in life. Generally, we are young enough so the pressures of cause and effect (karma) need to build a bit before we suffer the backlash. But the backlash from cause and effect will come. It's physics (action/reaction).

We give up the feeling of omnipotence only when faced with a force that we cannot control, cannot manipulate, a force that we must give in to. That is a scary situation. We face the fear of death (to a greater or lesser extent), because we must admit to ourselves that we have no choice but to ride with the force, like it or not. Remember me in the boat.

After a growth experience in which we resolve neurosis, we see life in a different way. We feel *real* for the first time, and we have increased self-assurance, self-esteem, self-esteem being the quality always lacking in neurosis. We see in this process a leap of faith, into an area of blackness, but only because of a strong opposing force. That force could be nature, daring do, a love relationship, a religious experience, even a psychologist. War experiences and conflict, as seen in many regions and affecting children, is another kettle of fish —mostly involving trauma and post traumatic problems. I am not writing about that, but it could become the biggest problem of all.

I write about resolution of neurosis extensively in two of my previous books: *When Mirrors Become Windows*, and *Personality and the Soul: Sixteen Women Show Us the Connection*. Below, I will list the ingredients to resolution of neurosis, but they remain the same as the transmutation process shown in the previous chapter.

1) The presence of an illusion of omnipotence, or a feeling of helplessness and indecision.

2) Faced with a strong force, we must admit vulnerability.

3) Because of frustration with how we are leading our life, we allow ourselves to be redirected by that force toward unfamiliar territory.

4) We pause, out of fear, and rethink: Do we really want to do this?

5) We then make a leap of faith into the unfamiliar territory.

6) We realize there was nothing to fear.

7) We view reality in a new way. Self-esteem enhances.

8) We attempt to convert others to our viewpoint. Ugh.

Such is resolution of neurosis, and we can begin to see how we can equate the leap of faith with the decision to go to ground zero and access soul qualities in personality integration. When we take the step to resolve neurosis, we are, indeed, accessing soul energies.

Let us go back to the process of personality integration

and see how it can directly apply to the concept of resolution of neurosis. It is here that we conflict with modern psychological thinking. In the classification of disorders, psychiatry and psychology separate neurosis from personality. They would place so-called *personality disorders* in a different grouping than *neurotic disorders*.

In dealing with personality, we realize that we live on a spectrum of qualities, and the worst, or even the best of the qualities, falls short of a soul quality. We live on the pleasure-pain spectrum, the spectrum of good and bad. Dissatisfied with that, we examine ourselves and we make a decision at some time in our life to do things differently.

Let us use the illusion of omnipotence as an example, since it holds such a key position in neurosis. We will use omnipotence as a personality characteristic and see where it leads us. As a personality trait, how does omnipotence play out in a personality spectrum?

If we put Omnipotence (the denial of fear) at one end of the spectrum, what should sit on the other end? At that end we see raw fear (called anxiety). How about Dread? The bookends of the spectrum will be Dread (the feeling of fear) at one end and Omnipotent (the denial of fear) at the other.

From Dread we can move to less fear with Timidity, thence to Ambivalence, and thence to no movement at all with Indecision. We can then only move to Decision. That marks the interlude. If we do not participate in the interlude then we stay on the personality spectrum and begin questioning our decision and move to Inflated, over-estimating ourselves, thence to Self-deception, which is hypocrisy, and thence to Omnipotence. We have the full spectrum:

Dread<>Timidity<>Ambivalence<>Indecision=0=Decision<>Inflated<>Self-deception<>Omnipotence

Dread is raw anxiety. In Ambivalence we see depression.

In Self-deception we see hypocrisy. In Omnipotence we see our present President (2018).

Between Indecision and Decision, we have the interlude, where a decision is made. That is the only point on the spectrum where a decision is made, and it is the beginning of something very important: taking responsibility, which is powered by energy from the Soul. Taking responsibility presents as the first step in dealing with karma, cause and effect. **Responsibility** is the soul quality.

By taking responsibility we face the possibility of consequence to our actions. We could also call this the first giant step in resolution of neurosis. It is also the key to resolution of adolescence. To take responsibility means that we must also take the consequences of responsibility. Herein lies the fear inherent in neurosis, the fear of the unknown, the bearing of unknown consequences. Here also we see the fear that keeps us on the personality spectrum, the same fear that turns us away from taking responsibility and going to ground zero.

By assuming the responsibility of power, rather than merely drifting in a sea of effects, we shudder just to think what might happen. We have no choice if we want to grow up. We know we must act, or else remain an adolescent the rest of our lives. We must go along with the force of maturation. Herein sits the force that we must give in to, go along with.

Conscious recognition of the soul (or soul qualities) forms the heart of the process of personality integration just as with resolution of neurosis. *Neurosis is just a word which denotes that we have stuck ourselves at some place on the personality spectrum, not being what we really are, as if the curtain for life had not yet risen.* Neurosis and personality spectrum are interlocked.

It is one thing for us to mutter that the soul probably exists. It is quite another to know that it does and to feel the effects of soul energy for the first time. At that point, as free and exhilarated that we might feel, we will soon learn that the work has just begun. We have made a great beginning, but we will need to face all of our personality characteristics, each on the spectrum, and move to the soul quality in each of them. Overcoming the delusion of omnipotence or the fearfulness of inaction, merely allows us to proceed. It looms as the first great step.

Aligning each of the problematic personality traits in our life becomes the work of our life, for a time, maybe a long time. As we make the decision with each trait and consciously cross the threshold of the interlude into the zero dimension of time, each time that happens, we take another step in overcoming our neuroses. We can define neurosis merely as the result of not entirely manifesting soul qualities. The problem resides within the personality which consists of barriers that we must deconstruct by making decisions to overcome fear (derived ultimately from the fear of death—as all fear is). As we overcome each personality barrier, we move closer and closer to the soul.

Psychologists will maintain that neurosis and disorder of personality are quite different. I would agree with that to some extent, only to the degree that people with so-called personality disorders go ahead and act out their personality characteristics, as dysfunctional as they may be. These people sit on the personality spectra, just like the rest of us. Perhaps they reside more off center, but we are all there, somewhere. These people do not try to cover up, hypocritically, their actions with neurotic shields. They go ahead and do what they want, and they make no excuses—though they might intentionally lie to cover their tracks.

Some people with personality disorders will have evolved to the extent that they have enough hypocrisy to erect neurotic defenses like everyone else. The neurotic defense, a form of hypocrisy, forms the next step closer to the soul. As dysfunctional as we might see hypocrisy, still, it allows the person a closer step toward ground zero. As we remember, it is at the stage of hypocrisy that we become particularly dangerous, a time when we can fool ourselves that our actions have the best of motives.

We all sit on the personality spectra. Wherever we are, we all have that in common. No one is better than anyone else in that regard. We all have defects. Some of us find ourselves at one end, and we like it there and make no excuses. Those are people with so-called personality disorders. Others of us want to move toward the center. Erecting neurotic shields allows us to do that since we have trouble taking the heat the closer we move toward the center. When we hit the center by an act of will, then we leave that spectrum and move on toward the soul, perhaps only to a more functional spectra, but at least advancing toward the soul.

We can see people who have severe (far to the extreme) disorders of personality—called character disorders by some—mostly with addiction today. To help them give up the addiction, we must introduce the neurotic shield. This does move the person toward the center, not to the center, but in that direction. To erect that shield, usually a connection is made with religion or some type or religious process, and that is perfectly fine; because as we have seen, such suffering people have stuck themselves far from center, and the addictive drug cements them there.

Later on, maybe, the addict can resolve the neurosis and make a giant step forward. Of course, one cannot make that giant step until the addictive behavior first stops. The addiction has served as a soothing balm so that he/she does not have to face the feelings and anxiety inherent in neurosis.

If we view our plight in life as one in which we all sit on personality spectra, then we have a conceptual framework that encompasses what modern psychology maintains as separate processes: neurosis, personality disorder, and character disorder. Change at all levels occurs through the process of transmutation. The changes we see are changes inherent in spiritual evolution. We do a person a disfavor if we label these stages or changes as disorders since these stages form an inherent part of the human condition, part of the psychological evolution of humans.

In esoteric thought what is called the ray of personality consists of those aspects of personality (the collections of energies) that any one individual retains. As the personality evolves through the resolution or aligning process, the energy ensouling or holding the characteristics together to make a coherent personality begins to become more consistently soul energy. Somewhere along the way, the energy becomes primarily soul, and the so-called personality ray becomes the soul ray.

Cultural Neurosis

We have treated neurosis as an individual thing, but how does neurosis play out in our culture? Does our culture (which *is* a collection of people) have neurosis, and if so, how might it affect each of us?

Since we have a capitalistic system of economics in our country, we might hypothesize that if neurosis exists in our society, then it might have a lot to do with *competition*. As capitalists we firmly believe that competition remains the best system going. Competition, it is said, produces the most efficient and productive in all areas. But who said that we were put on this earth to compete?

A biologist would shoulder me aside and whisper, "Competition is the way of the world. All animals and all plants compete in some way or another, and we are no different."

Maybe so. But a voice lurks in the back of my mind

which is saying that many of our problems have something to do with competition. We cannot merely dismiss competition as evil. We know that competition does produce results, and not all of the results are bad. Let us examine competition.

First, let us look at the dark side of competition. If I am competing, what can I do if I discover that I am losing in this competition?

1) I can cheat and lie and try to win that way.

2) I can try harder and explore other ways to win.

3) I can give up and

    a) become a loser and feel that I must be awfully dumb or inadequate, or

    b) I could erect an illusion of omnipotence and create all sorts of neurotic defenses to mollify myself that I am the person I imagine myself to be.

Basically, I can give up, or I can cheat, or try harder, or I can erect some sort of an illusion (shield) to protect myself. The shield results in neurotic defenses, and with these defenses I could claim that I would have won if only the other person were not cheating, or if I knew the right people, or if my arm didn't hurt, or if my uncle were rich. In any case, the other person (the winner) is either very lucky or taking unfair advantage. I can play the psychology game and insist that the other person wants to win because of some abnormality or improper motive.

I can even examine a chain of events leading up to the present and use this chain of events to prove my point. This is the paranoid defense. I could just as well use sour grapes, or I could intellectualize the whole thing and talk about it as if it were not me. I can readily see my own defects in other people, spot them right off, and I can despise those people for having those defects. Tricky.

Or I can bring out the defenses when I don't want to admit inadequacy; yet I do not want to cheat, but I do not want to try harder, either. I merely say that I am the unfortunate product of persecution, be that cultural, societal, genetic, parental, or personal. The problem resides outside of me. Oh hum.

Everywhere we see competition. We feel we must compete to fulfill our desires. The Easterners call desire the heart of human misery. Likely that is true. But in order to obtain what we desire, we must compete, because other people have desire and they want what we want, too. In the American culture, at least, desire boils down to competition.

When we compete, we can compete for fun, in which case we have the fun in the playing, not in the winning. We can compete for exercise to hone any skill, physical or mental. We can compete in earnest, in which case, competition can become an unfun job: competition to win, whatever the prize.

To win means being the best. If we want to win, then losing is not second best, it is doom and gloom. "I'm worthwhile if I'm a winner. I'm not worthwhile otherwise." By definition, when we see the desire to win, we see self-esteem problems; otherwise, we would have no problem in losing. If we wanted to compete merely to become our very best, regardless whether we win or not, then we would not see self-esteem problems.

We can see two aspects of competition:

1) Competition to win.

2) Competition to do ones best regardless of the outcome.

Unfortunately, our society rewards only people who win, though in recent years we see a movement to award every competitor as if everyone were a winner, but that really does not help much. The bottom line is that we have winners, and we have losers. The desire to win forms the dark side of competition. Though I do not have a great interest in the dark side of competition, I do feel that it reinforces so much neurosis in our society that we should know how it evolves.

Though we cannot blame society for our ills, we still have within our society certain built-in neuroses, as all societies have (Today, one of the biggest is that each person must feel like a star and have their own star-presentation on Facebook). These neuroses we all have to greater or lesser extent, but throughout the entire society these neuroses form the major stumbling blocks to soul access. We could say that competition in our society and the principle of our erecting our illusions of omnipotence (in the neurotic process) have a definite link. Each reflects the other.

At the beginning of the desire to win we see poor self-esteem. "I will like myself better, or my parents will be proud of me if I succeed."

Such statements do not sound so terrible, and they aren't, except for the neuroticism; and it leads to unhappiness, and discontent, even if the person wins. Winning may be the worst thing that could happen to the person. The person wins and discovers that she/he does not feel any better, hopefully learning from that.

If we want to win, whether we win or lose, we are in for trouble. When I speak of winning, I am talking of winning our desires. In seeking out our desires, we usually meet up with competition. The desires are manifold: a better job, more money, more attention, more popularity, more love. The list can go on and on.

Let us look at how competition evolves, and I am going to construct a flow chart for the evolution of competition. This will differ from the personality spectra, as we will not have the zero ground and soul qualities, etc., at least yet. We merely ask, how do we get from here to there. Hopefully, we will not end up with the situation that we find on some Maine roads, that you can't there from here.

We have, as a beginning, the primary motive for wanting to win: non-acceptance of ones-self as one sees oneself. We will call that "self-hate," though that may be a bit extreme for most, but let us use it as the extreme. From there we go to desire to win, or *competition*, in order to prove worth to self and others.

We have

Self-hate <> competition.

We have interest in the situation where the person does not win the desired goal. The person fails in attempts in life and does not win. We have

Self-hate <> competition <> poor performance

Feelings are next. The person feels anger. We could call this *jealousy*. We could also call it resentment, which is prolonged jealousy, but we'll call it jealousy.

We have

Self-hate <>competition <>poor performance <> jealousy.

What does the person experience now? You know, of course: the person faces *rejection*. Who wants to be around an angry, jealous, or resentful person?

We have

Self-hate <> competition <>poor performance<> jealousy <> rejection.

The person is rejected. She/he becomes a loser. If we feel like a loser, what do we do? We pull out the neurotic defenses:

"I never wanted it anyhow,"

"I'm too good for her, anyway,"

"He had it in for me all along,"

"This has driven me to drink,"

"What a bunch of snobs at that party."

We all can see these things in others, not so well in ourselves. We now have

Self-hate<>competition<>poor-performance <>jealousy > rejection > defenses.

The ultimate defense is *paranoia*, a mental state that we cannot refute to the satisfaction of the person feeling the paranoia; because that person can firmly maintain that certain events have led to this particular moment; and these events have been the doing of someone or something.

"All along you have been wanting to do this to me. Now I can see it all come together."

It is sad and frustrating (and sometimes scary) when we must face the paranoid defense in another person. We see it often in old people as they begin to lose their faculties, especially memory. In one sense the loss of memory breaks the chain-links in memory of events, and things happen to the person without the person remembering the causes of or the participants in the happening. The person is then left to his or her own devices to fill in the spaces, and many times the person draws upon paranoid material through fear.

A person may grow out of neurosis through the aging process. Many people do get better when they age. Others hang on to these neuroses, and when they begin to lose their abilities, their own admission of this loss remains too intolerable to their lifelong poor self-esteem. They deny their own limitations and fabricate a lie to themselves that they still have these faculties; and in that way they don't feel the hurt of further loss. They end up hurting themselves, many times leading to institutionalization and death.

Many people have spent an entire life guarding themselves. They are not about to give up just because they're getting old. If blankly, and pointedly, and persistently, someone faces them up to their deficiency, such as:

"Grampy, you know you can't drive anymore!"

Grampy will merely be pushed to the extreme defense of paranoia: "You are just trying to make me look crazy, so you can park me off in some nursing home!" Or Grampy might revert to the beginning of the spectra: "I must be the worst person in the whole world if I don't even know my own mind!" (Though Grammys more likely use this latter defense.) Either way, everyone loses. So why try to face them up to the reality? In most cases we will be forced into making the best of a difficult situation. Rarely easy.

We now have the final formulation of the evolution of self-hate and the results of poor self-esteem. We end with paranoia, something so far out that further talk does no good. We then resort to drugging our old people.

We now have the final pathway:

Self-hate <> competition <> poor performance <> jealousy <> rejection <> defenses <> paranoia.

Let us say that we had some insight, and we declared that it is time to get off this ride and do something different. Where would we begin?

We might consider doing something at the *competition* stage. We could win, in which case we would likely lose, because we then become even more self-absorbed; and we attach our self-esteem to more winning. All of that is neurotic.

If we lose in the competition stage, then we find ourselves at the *poor performance* stage. From there we can move forward, in which case, we feel anger or jealousy or some other negative feeling, and we immediately experience rejection from others because of the expression of those feelings. If these negative feelings get reinforced by, say, a supportive parent or a spouse, then we find ourselves in quicksand.

We create an ongoing problem if we attach our feelings of rejection to a thought in the form of a neurotic defense. Such is a thoughtform. If we do that, then whenever we approach a situation similar to that, the thoughtform will automatically appear. The more it appears, the more cemented it becomes into our reaction armamentarium. Do we have any other choices?

We could experience the rejection--that feeling of a loser. We do not like that pain. That is why we generally attach that feeling to the thoughtform: sour grapes, paranoia, depression, victimization, and others. But we want to do something different. So, let us try holding on to that feeling of a loser, as bad as we might feel. Let us say that we are experiencing the feeling of rejection or loss. What next? We have choices.

1) Become a loser. I am a loser. I am defective. I hate myself. Those attitudes merely move us back to the beginning, only we cement ourselves more at the beginning, and we may stay there in the reaction called *depression*. Who wants to do that? Unfortunately, a lot of people.

2) We can stop playing the game. Is this whole chain of competition *not* a game? When there is competition, then we have a game. Our need-to-compete forms the heart of neurosis. The one way out is to not play the game. *Merely stop playing.*

So, we stop. We walk out of the competition and say to ourselves: I am not playing anymore. I am going to sit down and watch, be a spectator for a time. This forms the first step into the unknown, and the critical leap of faith necessary in neurosis resolution. We have always been competing, usually secretly. Now we have stopped. What happens now?

We may find ourselves alone for a while. But if we watch, we will see where we can reenter the flow of events, and perhaps in an area that we might find satisfaction, even enjoyment. By stopping, we have brought ourselves near to ground zero on a spectrum, and we find ourselves at the interlude:

Nonparticipation=0=participation

From that point, we then make a conscious choice, our first conscious choice, since always before we had found ourselves carried away on the wings of competition. We then reenter our lives and find that with *us* making the decisions, we do not drive others away. The contrary is more the case. We drive people away with our competitiveness. We find *ourselves*, which turns out as a process of merely coming to ground zero and accessing a more soulful quality. What is that soul quality which reveals itself at the interlude of nonparticipation=0=participation?

It is the quality that resides between or beyond *not* being part of the process and *succumbing* to the competitive game. Hence, we are not withdrawn, nor are we overtaken. Yet we are involved. We are dedicated without being partisan. What is that quality? It is the beginning of *Wisdom*.

Wisdom comes from *detachment* to the degree that we can stop, look, and listen; then reenter. Life suddenly changes as if the other way was only an optical illusion, like an Escher work. The technique can take us to ground zero, and there we all have the potential to connect briefly with the incoming soul quality: wisdom.

Once we are making the choices, and we take responsibility, then we might want to compete, again. From then on, however, the game has changed. No longer does our self-esteem enwrap itself in winning. We look to self-improvement, honing skills, but not with the goal of winning. We may even attain a position of prominence. Either way, we cannot lose. Losing forms an essential part of the neurotic spectra. We win when we lose our self-absorption, and that is no loss.

Overcoming of the neurotic blocks of the illusion of omnipotence, cemented by competition, looms as the great task ahead for many of the people in our culture. When people from other cultures look at us, they see the results of these defects in our behavior. We, in turn, see the results of the defects ingrown in *their* culture. No culture has perfection.

For most people comfortable in the American culture, the moment of overcoming the illusion of omnipotence will be one of the big moments of that person's life. From that point we enter a path of *purification*, a path in which we learn to overcome some of the restraints of our own personality. Somewhere along that path, we will have an exposure to the full blast of the soul when conditions mix just right, and that blast will confirm to us that we are heading in the correct direction.

The orientation at the soul level focuses as one of *group consciousness*. We proceed from individualism, to groups, then group consciousness. Paradoxically, our interest in sports begins to give us some sense of group consciousness, though the concept involved is to win through competition. Even competition serves as a step toward what we will eventually find in the soul. We cannot throw out any aspect of human experience and merely label it bad. We see life in a continuum that leads directly to the soul. We all pass through these hallways, no one better than another.

Psychosis

If we were to experience high energy input, but we still resided on the personality spectra (and we did not possess secure personality integration) then all that energy coming into our system would produce a major misrepresentation. We call this major misrepresentation psychosis. We would misinterpret this energy on the basis of our maladaptive personality characteristics, and we could then suffer from any of the different types of psychosis—mostly schizophrenia.

A person suffering from schizophrenia accesses some energy from the soul and perhaps some energy from beyond the soul, but mostly from the soul. The personality does not have integration. The person may have a relatively normal childhood, but when the time comes when soul energy begins to manifest in life, such as at adolescence, or the first love, the energy lights them up inside like a Christmas tree with a short circuit. A fuse is blown. At that point, we begin to see biochemical damage that reflects what is going on in the mind. Some would say the propensity for damage comes first, inherited, then the behavior follows. Time will tell, but the important fact is that the person cannot handle the in-coming energy, and it becomes misrepresented.

The only known treatment so far is giving drugs that reduce the effect of incoming stimuli. This dullens the response of the personality enough that the person can live, even though the living remains painful. In schizophrenia, the energy of the soul energizes a dysfunctional personality; whereas in neurosis, a neurotic defense or a stubborn personality characteristic keeps the soul energy tempered and expressed as guilt or anxiety.

In neurosis, the person lives on the spectra of personality, and though there is some soul energy involved, e.g. being good is certainly closer to the soul than being bad; but good is not enough. Soul energy also manifests in reverse as anxiety, the sine qua non of neurosis, the fear of death. This anxiety moves us to find a way out. It that respect it serves a function.

In neurosis, the soul energizes a personality just as in psychosis, the only difference being how dysfunctional we might find the personality and the extent of the blast of soul energy in the schizophrenic. In the psychotically predisposed person, the personality is so unintegrated that the person cannot even use the neurotic shield. Soul energy merely energizes a poor circuitry and blows it out, and we see the distortions of reality seen in psychosis.

There is good reason to have compassion for people who suffer from schizophrenia. Compassion will allow these people to turn their lives around, perhaps not in this life, but later. Regardless how we might feel about the cause of schizophrenia, these people need compassion, compassion being an energy from the soul, the same source of energy that in their particular personality, for whatever developmental, or inherited, or previous-life reason, has blown them away. They need to see the quality of the soul, tempered, and carefully applied; and they can see that with our help.

The thing to remember about anyone with schizophrenia is that she/he is very real. The more real you are, the better you can communicate with that sufferer, and the greater your potential to help. If you are real, you can make an immediate connection. It takes no time at all. If you are not real, you are stumped. How many unreal mental health practitioners do you know? How many just deal out the pills? I don't know the answer to that question.

DESIRE

Someone once compared desires to sparks of motion. What does that mean?

If we look at desire, any one of the many desires that might pop into our consciousness each day, we could characterize them as little motivations for thought trips. Mostly, we cannot identify the instant when we begin to think of any particular subject, the very beginning, but we can catch ourselves somewhere along the line before we lapse into reverie or outright daydreaming.

Eastern thought maintains that we should not enslave ourselves to our desires. That makes sense. I think many Westerners would give lip service to that idea, but how many of us even think about desire? We all have desires—we might even consider them motivations—and I call these little motivations the sparks of motion.

When we talk about the little motivations, the little sparks of motion that start us to go in any particular direction in thought, then we can begin to think about ground zero, again, the center area of consciousness or that point between the two polarities of personality, the interlude. We know at that center point we find soul energy coming in. When we access that area, we have successfully transmuted the personality qualities on that particular personality spectrum into a higher value, one closer to the soul. We have added energy to our center of consciousness where the 'I' resides and transformed it, moving it into a (so-called) higher level of consciousness.

Motivation

What motivates us to do anything? Is it desire? That is a possibility, but we have not delved to the bottom of this matter. We need to look further. Can the motivation to do something come from outside ourselves? Let me use an example: You gesture to me, and I move toward you.

That is a clear example of a desire in me coming from a source outside myself and motivating me. You gesture, and I move. Let us look at that more closely.

We are not talking here of reflex action. Our reflexes are no different than animal reflexes, and we do not have control over them, like ducking when someone swings, the dog avoiding a swat. These reflexes relate to self-preservation.

How many of our actions each day link to self-preservation—life and death decisions? Maybe more than we realize. When we drive on the highway, we sit only a few feet from disaster. Every move we make is life and death. Frequently by reflex we maneuver our cars to safety. Put Thomas Jefferson next to you on a drive downtown, and he will be hiding under the dash. We have desensitized ourselves to what goes on around us, to the reflex behaviors that we have learned that preserve our lives every day.

You beckon to me, and I come; is that reflex? No, I had to make a decision in that encounter. I could have run away, maybe even fallen to my knees, or even stood on my head. I chose to come closer. Let us examine my move.

All of us have many choices for every move that we make. Different dimensions of time contain these possibilities in a time-stratified manner. I have made a decision on one possibility out of countless possibilities. The potential for all the possible reactions abides within me. I am somewhat limited, however, as I cannot fly; but I can do a lot of other things. Why did I choose what I did?

Whatever I choose to do, the idea behind it has already been thought before. The universe already exists. The thought behind all the possibilities of action already exists. The possibilities of all actions must reside within the realm of the Thought that abides within the structure of the universe in the first place. That is All Thought—the intelligence and organization of the universe. It is not chaotic, though that does not necessarily involve Deity, as we know.

In my decision to move toward you, I have first become aware of you and your beckoning (didn't your mother tell you not to do that?), and I have processed that incoming stimulus through my sensory apparatus and my energy bodies. Then I make a determination in my mental apparatus.

"Hmm, should I go or not?"

I might make my decision on the basis of desire. I figure things quickly, and if it seems like a positive move, then I come. If it seems like a negative thing for me, then I go the other way. Mentally and emotionally, I must clear the way for the action. What I do depends in large part on my personality and the state of my personality at that particular moment.

I will not come toward you if it is out of character for my personality. Maybe I hate to do anything another person wants me to do, especially if the person is you. I move in concert with my personality ray (my personal energy makeup), as the esoteric student would say. That ray includes all the possible ways of reacting contained within my personality structure. My personality, being what it is, restricts me to a great degree; and I ignore the multitude of other possibilities.

I still have choices within my personality. Given you, you being what you are, and my past relations with you, and given my personality structure and the emotional and mental state that I am presently in, does the combination of all of these factors leave any leeway to my action? Given all those variables in their particular configuration at that instant, will I always make the same decision and move toward you? The answer to that is, yes.

That particular situation, structured as it is in time, will never occur again. Never say never, but I feel confident that it probably will never occur again, especially since time is an element in the equation, and that particular time will not occur again. (We have a bit of a disadvantage because we are presently working on the first dimension of time, linear time.) I will always make the decision that I make at the time that I make the decision. No other choice. Crazy. You call that free will? Maybe not. Let us look further.

Choices

Lots of choices, but the basic choice is simple: yes or no. Whatever I do is an 'on' or 'off' decision, just as in computers. My personality limits my choices, but whatever action I choose to proceed with, I have responsibility for that action. An action, made up of multiple aspects of personality, proceeding because at that moment I choose 'on.' I choose 'on' because I felt positive about the decision.

Positive, what does that mean? Dealing with positive, or any gradation of anything, just means that we are living on personality spectra. Personality means polarity. Let us look at how a personality configuration might appear at the time of my decision, Figure (32), and let us only use four personality factors in this decision, though there would be more than that, even with simple decisions.

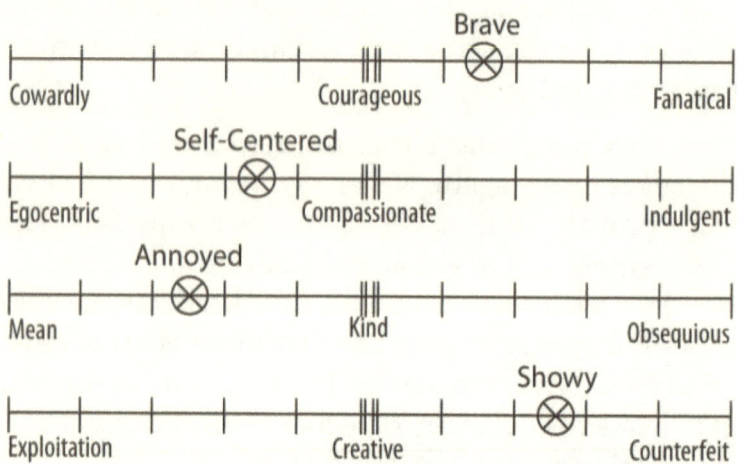

Figure (32) Portion of Personality Profile

The portion of a personality profile shown above only looks at four spectra. Where I reside on each spectrum is noted. We have four soul qualities: courage, compassion, kindness, and creativity. The extremes of each of the spectra are pretty gloomy. I am thankful I am not that far off center in relation to this particular activity, but I *am* off center.

I do what I do (noted by a circled 'X'), because (1) I am sort of brave, because you have the reputation for putting people down (2) I am self-centered because you owe me money (3) I am annoyed, because I am hungry, and (4) Others are watching, and so I will be a bit showy and put on a good face.

All in all, everything adds up to the result that I move toward you. I make that decision in concert with my personality, and my personality integration profile remains about the same. If I cannot hold the personality characteristics (my emotions mostly) in their present configuration, then what I do will vary.

"I'm sorry, but I have to go," I might say, and then walk away, feigning important business to attend to.

Decisions that allow us to move easily in concert with our personalities we find easy as punch. Those decisions in which we must reevaluate ourselves and perhaps change in order to make the requested move, we find those decisions difficult. How does all of this help us with desire?

Describing Desire

Desire is the attractive feeling that steers us to seek out certain activities in our lives that reinforce our personality configuration at any particular time. What we seek is food, basically, for the (personality) animal inside us. That animal has the configuration of our personality and it seeks food so it can continue to exist as is. Without the reinforcement of the food —getting what is desired, or even the fantasy of getting it— the animal will die. By death I mean the particular personality configuration will die out, only to be replaced by another one that needs food.

Desire has very positive aspects about it evolutionarily if we consider humans as animals. An animal seeks food, shelter, comfort, sex, and some companionship. Desire is great for reinforcing the personality characteristics concerning obtaining those things from life. The problem appears when a human wants more than that and realizes that life consists of more than just applying animal principles to life. Desire at that time serves as an obstacle.

We cannot, however, throw away the fact that we are human. We still have this living animal within us that wants gratification of animal desires. We have to live with that. We must accept that.

Going back to our center of consciousness that makes decisions, how do we free ourselves from the enslaving desires? When a desire pops into our head, do we then merely suppress it, and then try to move on? In other chapters we have talked about suppression only meeting with further resistance from the animal within us. We must, therefore, seek out another method. We must use the *will*.

The word "will" refers to decisions that we consciously make when we reside at our center of consciousness. We then are talking, again, about the transmutation process. We bring ourselves to ground zero, and we instantaneously hold ourselves in brief abeyance and access the soul quality. Then whatever we do will conform to a greater degree with the soul. The esotericist would say that the modification of the mind stuff must instantaneously meet the Divine Will. That is their way of saying All Will or all the Will energy that exists.

What exactly does that mean when we are referring to desire? We asked the question earlier if it were possible to have any of our actions motivated from outside ourselves. Perhaps we agreed that certain reflex actions might qualify for that. Allow me to put another twist in this adventure.

The will to do anything, I mean anything. Where does that come from? We might reply, "From our desires." We see something that we desire and we seek it. Our desires base themselves on our personality. Okay. But the will to make the decision to do anything, even that on/off decision that we must make in accordance with our personality each time we do something, where does that come from?

Universal Purpose and the Ego

I might get a bit metaphysical here and say that the Will of the universe exists every bit as does the Thought behind the universe. The universe exists, and it reflects the thought behind it. All possible thought currently exists in the creations. That same line of thinking we can apply to Will. The Will behind the Thought behind the creations exists, always did, always will. I say 'creations' as the universe has an origin. Call it what you want. But the fact that everything hangs together with thought does not mean that it has to be somebody's thought (perhaps I should say, Somebody's).

Any will in the universe must reflect that Universal Will. So, we say solemnly, "It is God's Will." Who could disagree with that? I am sure that some do. And some of *you* are reading this. The character of matter in the universe and the structured physics in the universe all point to intelligence. It is, indeed, inherent for matter to proceed as it is proceeding through the characteristics of energy—that is what I am calling Will. Who could disagree with that?

The will to do anything in each of us partakes of that Universal Will, but we do not have all that Will. We only have the will necessary for us as persons. We access that energy of will in a purer form, with less distortion, as we approach our center of consciousness, and in a more concentrated form as we increase our level of consciousness. The soul remains our personal energy source.

Let me insert something that we will go into further in later chapters. *Individuality is a quantum of Purpose.* The center of consciousness, which I am calling the 'I' or ego, is actually a very small portion of Purpose, the same Purpose or energy that created the universe. The substance of the ego is energized by Purpose. This is the closest we can come to Purpose (of the universe), but that's pretty close: it's us. Not to say that just because our ego is energized by Purpose that we will always do things in concert with Purpose. We might or we might not, depending on how we use the tool of the mind.

We are now talking about Purpose, and Purpose (as was said in the chapter on Plan) precedes Will. The will necessary for us to seek out what we desire receives its energy from our center—we might call that our Divine spark or essential spark, whatever—but we generally use that energy to seek self-serving desires. I realize all of this sounds foggy, but there's a later chapter on Purpose, and hopefully I can better clarify what I am saying. It is not easy.

Our Responsibility

Incoming stimuli, like the sight of someone else's new car, is not the cause of our desiring a new car. That incoming stimulus consists only of a sensory image. When that sensory image happens to stir up waves because the soil is fertile, the personality configuration being such that the stimulus resonates with a personality that trends toward "showy," then a little desire storm sets to twirling. The thoughtform stimulated by the tempest of the experience, excites a response. A response from where? We, ourselves have to respond. We allow the reaction to proceed or not.

From our centers, we still have to activate, 'on,' or 'off,' but if we go for 'on,' then we unleash this energy to proceed to a goal as if it had a mind of its own, which it does not. The brief touch of our center to get permission is so brief that we have no consciousness of it. Then the thoughtform related to our personality, however it might be configured at that moment, determines the action.

"I guess I sort of lost my head with the insurance agent when I bought the three-million-dollar disability policy for my canary."

The center of consciousness said, "On," and the thoughtforms reinforced by the insurance agent became energized enough that they were reacting while the center of consciousness sat back murmuring, "Oops."

Every time we make an emotional decision, we still have the responsibility for that reaction, and we hold the blame for abrogating the subsequent reaction to something other than our centers. We gave permission, even if we do not remember doing so.

By saying, "Yes," to the force of personality, we gain experience, and we set karma into effect. We learn from the consequences. By saying "No," or "Stop," then we can come to our center—without a specific reaction in mind—and we open ourselves to the soul's suggested energy input, a better, more humane, more evolved response:

"No! My canary doesn't need such a policy, I'm going to invest my money in speed bumps."

This small bit of enlightenment comes to us, like intuition, immediately and without volition. The soul eventually wins out, either way, given enough time. We have to be paying attention.

We call the input that we receive from the soul during this process, transmutation of an idea, and we register it mentally. The soul is the true motivator for our physical lives, and beyond the soul we can hypothesize another center of will, and other more inclusive centers, until we have encompassed the entire universe and involved the Thought and Will (and Purpose) of the creation of the universe, all related, part and parcel to our little wills and little thoughts. That's a mouthful, but All *will* derives from *All Will*.

Let us now go back to where we were when you gestured to me, and I moved toward you. I made the decision to move, but it makes a big difference in the encounter whether I was allowing the personality to control, or whether I had the ability to access soul energy with that decision.

Many times in life we have the need to make decisions immediately. Whatever I decide to do, I still make the decision to do it, whether it is run to you, or walk over to you and say, "What's going on?" Integration of personality through the transmutation process means integration to access soul energy whether we are in a hurry or not, and it means all the time. Ugh! You mean ALL the time? Well, that's the goal.

Sparks as Ideas

I made the statement at the beginning that desires were like sparks of motion. That statement begins to come into better focus. Accessing soul energy through transmutation, we expose ourselves to *sparks of ideas*, the same ideas that reside within the Thought of the entire universe. It is as if the universe were filled with stars and each star were an idea, and we learn about the ideas one at a time, sometimes in bunches when we can make associations, and sometimes in a rainstorm of ideas, which in itself is a pretty amazing thing.

These sparks actually motivate us, as intentioned as they might appear, but our personalities filter this energy as we react far from center on our personality spectra. We use the energy coming from the soul for things like consumption, comfort, need, fantasy, all of which are affected by the types of stimuli that are coming in from our everyday lives—mostly television commercials (or nowadays—material coming over the internet). In reality, the sparks of our motions consist of idea-sparks and those sparks, if looked into, reveal the true identity behind whatever we are reacting to at the time. Something to think about.

It is the same with the arts. Dance reveals the true identity of human motion. Singing reveals the identity behind voice. Poetry reveals the art of speech. Art, in general, discovers the identity behind whatever the artist encounters. The art in life is the soul in life and occasionally more than that.

Ideas

Our culture evolves and changes based on ideas. History is nothing but the evolution of ideas, and we find that ideas proceed to greater and greater abstraction, that is, toward more inclusiveness as time goes by. We see this all the time, but we fail to recognize it. An example would be *to watch out only for yourself*—an animal-like reaction. This was the standard up until about 2500 years ago (in western civilization) when a note of altruism began to show itself within humanity, seen in the Greek philosophers and later within the Christian Church. In the New Testament, we read "To love your neighbor as yourself." As we come to the 20th century (now in the

21st with this rewrite), this idea has expanded to include loving all of humanity (humanitarianism) and to love the earth (stewardship), as the idea becomes more and more inclusive.

What we call creativity is a means for accessing ideas. When an idea first introduces in a culture, it takes a while to have a significant effect. It must sift through, and over time nearly everyone accepts it. Then it becomes part of the culture, and we don't even think about it anymore. *All men are created equal.* That idea is a given and intrinsic in all thought for some people, but it is still evolving in some circles. Someday everyone will automatically accept that idea.

Desires come from ideas. Ideas are eternal, and they represent what we have called Plan, the blueprint for possibilities within Creation. We can only desire what is possible. How could we know what cannot happen? It has never happened before. Yet we are surrounded now by all the ideas of the universe, only we don't have the eyes to see them, even though they are staring us in the face right now. Ideas appear when we are ready to realize them.

All desires, no matter how perverted, have some element of Divine or Ultimate Will or Purpose within them since the one who chooses to desire is *us*, and the us-ness or we-ness that makes each of us an individual partakes of the Divine Spark or the spark of identity, defined as an element of Divine Purpose (Universal Purpose). A bit of the Divine Spark exists in everyone, and all our motivations that we have every day reflect that Divine Spark in some big or little way. You materialists, you know what I mean, but I have to stop explaining it each time, since this book is too long already.

The task in life is to better and better discriminate this energy that motivates us, to better refine that energy to reflect the idea or ideas that reside behind motivation. In that sense, our desires become transmuted to greater and greater abstraction and inclusiveness. This is just another way of saying that we better reflect higher and higher levels of consciousness such as the soul, and even beyond that.

To understand others, to appreciate others, and to have compassion for others, remains the art of recognizing the ideas behind their desires. The truly great leaders of our world have that capacity, and we can see why people want to follow them, because they feel appreciated. They know that their leader can see the good within them, however great or small that might be. There is so much going on in all our lives that we all can relate to, in each of us. Lately we have focused on politics and allowed it to divide us. But we each are humans. We eat, work, play, have common problems and perplexions (not an official word), and we have so much in common. Too bad we have to focus on the divisions of politics.

This chapter about desire sets the stage for the next chapter, on Will.

Fasten your seat belts.

WILL AND SPACE

We are going on a wild ride through space, a mental Space Mountain as we attempt to answer the question: How do things move through space? It's not easy.

Space has structure, though we do not know exactly how it is organized. According to quantum theory, when anything is set into motion on earth, as throwing a ball in the air, the ball returns to earth only because the earth distorts space/time. The structure of space likely changes because of matter within space, and there is a significant possibility that the quantum entanglement of particles creates what we experience as three-dimensional space. Further, when we measure anything in space, the parameters of space/time change, and the tangible properties of substance appear, whereas before, they were only probabilities. If exactly true, then humans are a *necessary* component of the physical world.

What I just said is a few sentence summary of some of the important quantum theories since Einstein, not that you need to understand what I just said. Yet, those kinds of considerations are an apt introduction to what I am about to present to you, making what I am going to say appear less fantastic (I hope). But it *is* fantastic in view of our present knowledge. I can only say that this material came from the same source as this entire book, but I cannot say that I understand it, though it does possess enough logic that I felt I must present it. Here goes. I am referring to the block, as you will see.

We could say that matter can (potentially) fill space

reserved for matter. We could term such areas of space as potential matter-in-space. Potential matter-in-space would contain the products of thought, much as a block of marble will contain the idea of the sculptor. Thoughts become things in potential matter-in-space.

Only certain combinations of matter can fit into potential matter-in-space, as these combinations must have congruence with Universal Thought, or the Thought of that Entity that created the universe, or that thought or organization behind reality (Take your choice.). The Thought behind the creation of the universe must contain all the possibilities of matter and its combinations. Through science we discover these secrets of the universe. We do not create any new ideas. The ideas we have exist already as reality, including all future ideas.

Matter arranges in space, coming together into larger forms, from molecules to galaxies, according to potentials already residing in space based on the organization of what I am calling Thought (for want of a better word) which produces the possibilities, which includes the processes of physics. We cannot force matter into forms that do not conform to potential matter-in-space. Potential matter-in-space, as a term, describes the possibilities within any area of space, and likely depends on the unknown structure of space.

A cloud of hydrogen gas and a lighted match would serve as an example of an incongruency within an area of space. An explosion results, with traces of water as a residue. We therefore see the connection between potential matter-in-space and Mind. As something is created, wherever that something goes, Mind, though Thought, has prepared potential matter-in-space for the possibility or impossibility of its coming or going. It's all physics.

Behind all thought resides Will. When something moves in space, that which prepares space for this movement is Will, though as mentioned above, thought (or Thought when speaking of universals) governs the possibilities. We could say that everything moves into preconditioned spaces. Whatever it is, it will move only into one particular area of space, because that is what is going to happen and Can happen.

Chaos and Prediction

Take a handful of sand and throw it into the air. The individual grains of sand will land in particular areas according to how they were positioned in the hand, the force of the throw, and the direction and force of the air currents, and other factors. We could not predict where each of these sand grains would land, because the system has many complications. But where the sand grains will land *is* predictable, if only we had the ability to do such prediction. Given all that goes into the throwing of the handful of sand, these sand bits will land precisely where they are going to land, according to known physical principles, but in such a complicated way that we cannot predict. These sand grains could not have landed anywhere else than where they did land. Some researchers in the topic of chaos maintain that certain systems are beyond prediction. Again, we are dealing with *effects* and not *causes*, when we should be asking the question: What are the causes behind our physical laws, and which are the effects?

The potential for all that precise movement and the precise final arrangement of the thrown sand was already present—or else there would have been no sand in the air, no sand landing anywhere, no such movement at all. The complication of systems prevents us from predicting such outcomes beforehand, but if we could, we could predict earthquakes and so-called accidents. As yet, we cannot.

A Child Moves a Block

The will involved in any system must prepare potential matter-in-space for activation into matter as we know it. I am stating this as a general principle. Let us look at an example: a child's toy block.

Thought of someone has created this block sitting at this particular moment at this particular place in space. It was possible to create such a block, obviously, because it was created. That means that it sits where it sits because of congruency with the possibilities of matter-in-space. The Thought of the Creator (or the Organization behind the physical laws of the universe) created the possibility of that block existing as it does. The mind of a human constructed the block.

The block sits on the floor. The child pushes the block with a finger across the floor. Once the block begins to move, *potential* matter-in-space changes to *actual* matter in space, which appears as movement of the block. This has analogy to the question, does fog move; or do the conditions that make fog move?

I will say that the conditions move that make the block exist with movement (though I do NOT know if that is actually true). That movement depends on *will*. The block moves because of the force of the child's hand, a force that receives its energy from whatever created the universe in the first place —One Will, *all will* actualizing from the One Will, individual or natural. We can say that Will (or will, a quantum of Will) causes potential matter-in-space to actualize into matter as we know it and causes the phenomenon of movement. What?? Where the block sits is where it exists. Where it was a moment before, it no longer exists there; its existence or *being* there, has disappeared. More truly, where it was, the space has reverted to potential matter-in-space.

All movement is a blur. We cannot see, actually, how matter comes in and out of existence as it moves through space, even when it appears sitting perfectly still. All matter is in movement, the molecules and the atoms. Space has meaning only as a potential place for matter. Will prepares potential matter-in-space and allows it to actualize into something created by thought.

If we go to the *Yoga Sutras* of Patanjali we discover a discussion of these subjects. We find of interest the following declarations:

"I am." The 'I' identifies with the changing form in movement, the blur. FORM.

"I am That." The 'I' identifies with the thought that creates the object in question. The thought is the thing, space not regarded. THOUGHT.

"I am That I am." The 'I' identifies with the Will, the element that causes all the movement independent of our time dimension. WILL.

We have the following formulation, according to esoteric thought:

ego > Soul > Ego

I am > I am That > I am That I am

The child pushes the block, Figure (33). Space is met by all the surfaces of the block as it moves along.

As the block advances, Figure (34), potential matter-in-space in the area around the block activates to form the block (noted as A). At the rear of the block, matter deactivates to revert to potential matter-in-space (noted as B).

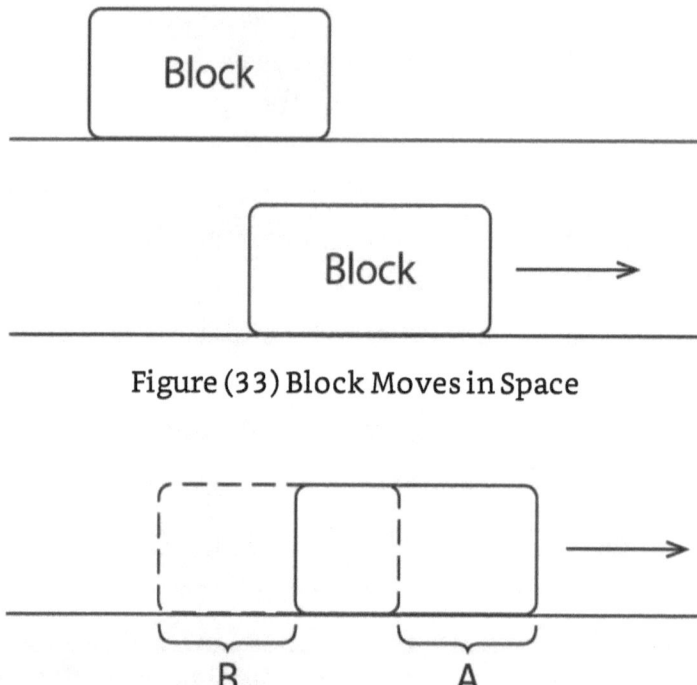

Figure (33) Block Moves in Space

Figure (34) Potential Matter-in-Space is Activated

Movement: Creation and De-Creation of Matter

By now, you probably are beginning to realize what I am saying: The so-called moving block does not move, but it does change location. When subjected to the force of the child's hand, the block continually creates and de-creates such that it changes location. Will moves forward, along with the block, through the intention of the child, and will creates and de-creates matter as it travels. We call that *movement*. Using the analogy of the fog, the *conditions* that create and de-create the block advance through space.

Because of the movement, we can imagine a curtain opening to either expose matter or expose potential matter-in-space, Figure (35).

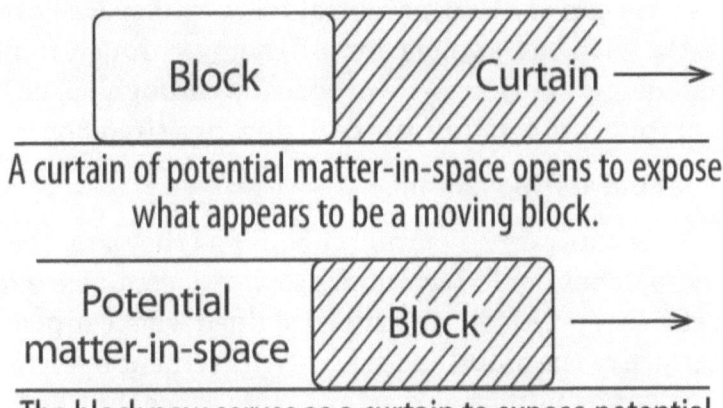

A curtain of potential matter-in-space opens to expose what appears to be a moving block.

The block now serves as a curtain to expose potential matter-in-space (blank space to us) as the block passes out of existence as it proceeds forward.

Figure (35) The Curtain in Space

The edge of the block, both front and back, in the direction of the movement, is like a curtain or interface between matter and potential matter-in-space. The curtain begins to open when Will is applied to the block through force, the force of the child's finger. We could say the same if the force were a natural one such as gravity and the block was falling instead of being pushed. Space/time accommodates to motion.

The child applies force to the block, and as that force causes movement, the block interfaces with potential matter-in-space in the direction of that movement. As the child pushes the block, what happens to the force of the push while the block moves?

First, we have the will of the child (to push the block). Will transfers into force with the push. The block begins to move because of the additional energy applied. That energy is movement. *Movement and energy are the same.* The amount of energy that the block contains is a function of how fast the child pushes the block. We now have

Will > Force > Energy (Movement).

We can say that potential matter-in-space converts to matter with the application of energy through movement. The energy continues to propagate through space until it peters out through friction, collision, or a tired finger.

Soul and Movement

As contrary to common belief as they are, these ideas of movement could have some sort of acceptance except for one further problem: We have explained what can occur at the interface of the block and space as movement opens a curtain of matter or potential matter across space, but how can we explain what is happening with, say, the middle of the block? The middle of the block does not interface with the space through which the block travels.

How can the matter of the middle of the block continue to change as the block seems to move across our field of vision? If movement is only potential matter-in-space changing to matter and back again, the entire bit of matter involved, the block, the entire block, must revise as a unit, while the block changes location. How could we possibly explain that?

At this point we must talk of the concept of *soul*. With the block we are dealing with an integral unit. It has been ensouled as a block through the thought involved in its creation and the work involved to make it. The block will continue as a block unless changed through the application of more energy in some way. The block, the entire block, encodes as a block. If not, it would not be a block.

Any part of an integral whole represents the entire whole. The soul of a part is the same as the soul of the whole. This concept of soul allows us to understand that all the atoms of the block have *block-encoding*, and when even one atom begins to change because of movement, then the entire block must simultaneously change, too.

Movement is the instantaneous transformation of an ensouling code through an area of space by the action of will, activating potential matter-in-space into matter in direct relationship to the force applied. True? It sounds good, really intellectual. If it *is* true, we have a long way to go in order to prove it.

Will and Movement

How can matter be created so easily? We know that Einstein's equation $E = mc^2$ states that energy can convert to matter only within the factor of the speed of light (c) squared, a considerable amount of energy. We have to realize that the energy behind the push of the block is will (a little portion of Will). That which changed the block's location was will. We consider Einstein's equation:

Energy = Matter x speed of light (squared)

We can substitute in that equation: Movement for energy (as we are saying that movement and energy are one), The Block for matter, and Will for the speed of light squared. We then have

Movement (change of location) = The Block x Will.

Does it mean that Will equates to the speed of light squared? No, it only means that movement relates to how much Will we apply to the block. The maximum amount of Will that can be applied for movement for the maximum amount of matter (the universe) is c^2. At that level, we find universal creation level of energy—a really Big Movement (hopefully this is not a reference to the material here presented).

The impetus behind movement is Will. Will, transcribed into movement, changes potential matter-in-space into matter. Will, itself, is potentially c², but in actuality, *it always equals the job it happens to be doing.* All Will must at least equate to c². The will of the child partakes of All Will when it applies to what is needed for the job.

Evolution of Matter and Consciousness

We have two simultaneous processes:

1) The evolution of physical matter and the energy involved in that, and

2) The evolution of consciousness.

Einstein's relationship governs both. Evolution of physical matter equates to the physical energy in the atom. The energy of the evolution of consciousness relates to the energy of Will. In Figure (36) we can see how the evolution of the physical and the evolution of consciousness relate.

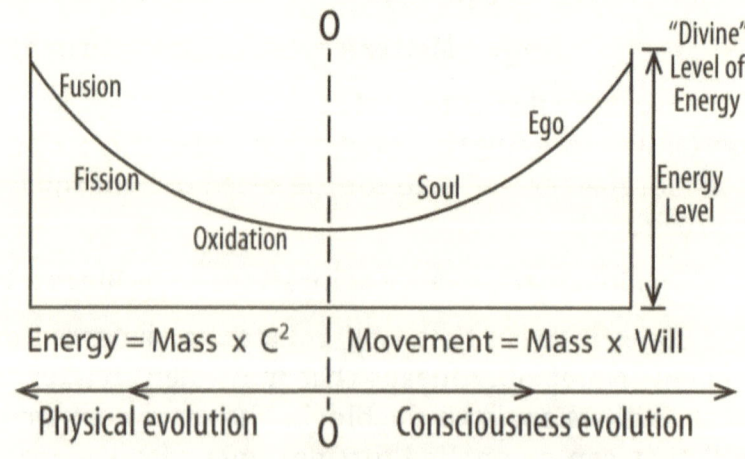

Figure (36) Evolution of Form and Consciousness

In the diagram, both processes have ground zero in common, and the energy levels increase proportionally as physical energy on the left of the diagram and energy of consciousness on the right side of the diagram. The energy levels, however, are the same in either system, both approaching as their limits, the Divine (or Universal) level of energy.

Concerning the phenomenon of movement, matter is created through movement across an area of space, and the energy of creation appears to be at low levels—just enough will to 'move' the object. We cannot measure will, itself, but I am saying that it has the power to convert potential matter-in-space to matter by moving that matter.

Powers of Will

We do not understand will. We fool ourselves when we look at the power we see in physical systems, such as the power within the atom, and we assume that atomic power dwarfs anything inside ourselves. Whoever said that faith could move mountains knew something about energy. Faith requires the will to believe, and yet, *will* far exceeds faith in power, though faith allows us to approach areas of reality beyond our powers of consciousness. Theoretically, we have unlimited power as a potential, but that power has limit now because of certain fail-safe mechanisms built into the system.

The movement that we sense is the blur of the movement. It adds up to a blur of the past, present, and the future.

Past = deactivated matter.

Present = matter activating

Future = potential matter-in-space about to activate.

What we sense as Now is the blur of the process. The actual Now, or the Eternal Now, is a function of Being, which is the reality of the object when it was created as an object. Being has nothing to do with movement, as the reality of the object resides beyond any time and space parameters.

Will is the process, the entire process, or that which sets the entire process into movement: the 'Hand' of the Creator, so-to-speak. The realization of Will concerns the dynamics of movement or the activation and deactivation of space to contain matter already created (if that is true). The matter goes 'in' and 'out' of existence as it moves, but it still 'is,' because it had been created. That 'is' is the object divorced from movement and consists of the reality of the object existing within the Eternal Now, which is a function of being and ultimately, Being.

We have Einstein's equation: $E = M \times c^2$

We converted the equation to: Movement = Matter (the block) x Will

We transpose that equation: Will = Movement / Matter (Movement divided by Matter)

This transposed equation represents the perception of Will. The power of Will has responsibility for movement (and matter) and is governed by Einstein's relationship (relativity) because once something is created (at its beginning) it is relative to other things then in existence. With Einstein we are talking about movement. What we see in movement relates to the ultimate extension of the powers of consciousness (Will in this case), and we can describe that power in terms of c (the speed of light). We cannot describe Being; yet it remains our ultimate reality, and it is beyond any parameters of space and time that we know. I know all this is complicated, and I am doing the best I can to explain it.

Reality

How can this child's block be its reality? Really. We are so used to sensing things about us, and now I am saying that what we perceive is not the reality at all, only the blur of the reality. The true reality exists beyond time and space and consists of the Eternal Now. If what I sense is not the reality, then neither are the senses. The senses only respond to the blur. We respond only to appearance, not to reality. Hmm...

The senses pick up traces of the physical object, the block, its visual appearance via our eyes, its feel through our touch, its smell through our nose, and its taste, if we are the child who chews on it. These are merely traces of what the block is, physically. So, what is the block?

If we only look at the dense matter aspect of the block, then we would have to turn, once again, to Patanjali and the *Yoga Sutras*. In that text, he discusses the aspects of matter and how we become embroiled in these aspects through attachment.

While I am writing this, I am in an airplane, and I see a man move from his seat, pick a magazine from the magazine rack and return to his seat. I can respond to what I see by making analogies to other experiences I have had in the past. I can respond to the appearance of the man, his race, his dress, his cleanliness, and I can respond to his movements. Did he move in a smooth and athletic manner or disjointedly and clumsy? If he appears spry and smooth, I might feel envy, awe, maybe respect. If he appears clumsy, I might feel pity or disgust. I might respond to the magazine he chose. Who would read a magazine like that?

All that we see, Patanjali would say, results from the Divine Will (One Will, or the energy of Will) working upon matter in some way through our consciousness as humans. We have choices as to how we either react or not react to what we sense. Usually we respond with thoughtforms, old thoughts attached to emotions, and we go no further than that. Patanjali would say that we imprison ourselves by the gunas, or the qualities of matter. True enough.

What other choices do we have? We know that energy governs all that we sense. Concerning the block once again, the reality of the block is the energy making up the block. That includes all the thought, the labor, and the history, that went into this block, including the materials. We could call the aggregate of all this energy the ensouling energy of the block. If that is a reality, then what senses that?

The mind must serve as the sixth sense and have the ability to sense this reality of the soul. If the mind is the sense organ, then what are the sensations being sent out by the soul that the mind can sense?

Plato has a lot to say about all this. Suffice to say that he thought that the *mind* senses qualities of the soul, just as the *physical senses* sense the physical qualities of the block. The primary sensations that we receive from the soul revolve around beauty and truth. Beauty is not appearance, though appearance can mirror beauty. Beauty comes from within and displays the soul. Beauty involves many subordinate qualities, all coming together to form what we call the beautiful: goodness, power, utility, harmony. With the realization of these qualities we have a sense of appreciation for truth. The mind as a sense organ allows us to sense the constant and abiding reality behind the changing or illusory appearance of the physical object, in this case, the block. Time for a pause? I feel it.

Beyond the soul of any object or organism, the mind has the capability to sense the Will. This Will is the power behind all manifestations, all objects, all thoughts, all love. It is the power behind what we sense in the soul: the soul qualities. We can term the qualities (the sensations from the soul) as the Spiritual reality. We can term the Will as the Divine (or Universal) reality behind the soul.

The Will of the Creation translates into the will, and beauty, and good within the soul. At the physical level it translates into goodness, goodwill, and what we call beauty (not what beauty contests call beauty, though they try). What I am saying is that in order to access the goodness of the soul, we must *will* it. That is the Will-to-Good. We cannot sit passively and hope to see that. It takes effort on our part, such as in personality integration.

Once we can exercise the Will-to-Good, and we translate it into our own lives as goodness and goodwill, then we also find ourselves accessing the Will of God (or the Universal level of Will), an immediate power that some call intuition. The Will-to-Good, goodness, and the power of intuition work together as our own personal program of salvation and spiritual evolution.

If we can truly sense or "see" the block, then we can see not only the physical qualities through our senses, but we can sense through the mind, the soul of the block and the Divine (or Universal) aspect of the block. An artist, with the eye of an artist, senses the block with the mind, and the resultant work of art intuitively demonstrates some aspect of the reality of the soul or even a bit of the Divine aspect of the block, in some abstract way. Through the creative act (including the creativity of everyday life) we can sense the Spiritual and even the Divine (or Universal) aspects of our surroundings.

I believe that in the distant future, we will evolve to the point that we can learn to recognize and channel Will, and once we can do that, then we will have the conscious use of the power behind movement. Space travel would then be possible and not limited by the speed of light, but by the speed of the mind. We could not be entrusted with that ability now, other than to push along blocks. Quantum entanglement would become a reality to us, independent of the speed of light—maybe equal to the speed of light squared, but that is pure supposition.

So, what is the block? The reality of the block is what it is divorced from movement (time and space dependency) and dependent upon the energy of all the forces of its creation which is a function, ultimately, of Being. When we connect ourselves to the Eternal Now, we sense the beingness of the block, the soul energy of the block. Allow me to go over again the perception of the block.

Any object contains many energies. In the chapter on Ownership, I defined the energies in a paper cup in relation to earth, air, fire, water. I defined those energies in modern terms. The reality of any object is all those energies.

Okay, what do we sense?

Artists attempt to portray realities underlying appearance. That's easy to say, but again, what do we sense with the object itself?

That experience is on a spectrum of realization. At the elementary level we appreciate what it looks like. Does it please us to look at it? Thence, more abstract, we can see how its function mirrors its appearance (the art of utility), and thence, more abstractly, to the mind of the creator (person or not), thence to how it fits with other objects (its utility environment), and thence to purpose (in the case of the block how it contributes to play and learning), and thence to the kindness of its very idea—the motive behind its creation, thence to the will behind the motive. The kindness is what I was referring above as the will-to-good. That is a rather elementary exposure to will.

From this last step we enter the universe of the will, much beyond the finger pushing the block. We will learn to channel will (Raja Yoga), but to have great effect, our motives must be of the purest, like perfect. That's the key to the power of the will, which in its purest form we call holy (no other word for it). That will is also termed Spirit, but when applied to matter, it is called force. The Energy when applied to matter to produce force, just that energy, not the matter, that is E, or energy-in-itself. That connects to the Eternal Now. I hope this is making sense to you. The next two chapters will help.

Esoteric Physics

We have talked in this chapter and in past chapters

about the physics behind the physics that we observe. I am saying that what we call physics consists of *effects*, as opposed to the physics behind our physics, which involves *causes*. The universe only allows certain types of reactions to occur. Those limits set the boundaries for our physics. I call those limits, which are beyond our mathematics to describe, esoteric physics: the physics of energy-in-itself, the physics of Divinity--being defined (if you want Divinity) as Energy-in-itself, divorced from matter. Divinity is divorced? Well, we all make mistakes.

Space is immutable, and space limits all reactions that occur within it. Space is the ultimate agent of karma, since it causes all reactions to go in certain prearranged directions, the only directions possible within the parameters set by the physics of space. *Space is the container of matter and structures all changes of matter.*

We spoke earlier about the evolution of matter and the evolution of consciousness. What is the container for consciousness, just as space is the container for matter? When something comes into existence (as matter) then we are dealing with the phenomenon of *creativity*. With creativity we have *sound. Sound is the container for consciousness just as space is the container for matter.* Mind uses the process of creativity to express the possibilities of sound. We remember that physicists can, with their instruments, 'hear' the sound remnants of the creation of the universe. *The limits set by space and sound involving matter and consciousness give the parameters within which we exist.*

The one point that I want to make is that space is timeless, and it sets the boundaries for all reactions within creation. Within space (within Creation) we find the evolution of matter and consciousness, and that involves time and direction. So, we have two parameters: The first is space, which is timeless—as immediate as immediate can be—and it sets all the boundaries for all reactions. The other concerns matter and consciousness and that involves time and evolution. It is at this point that we are beginning to see the components to the debate between the *immediacy of creation*, which some Biblicists maintain as truth, and the *time dependent* process of evolution, which science believes. I will make some important statements concerning this later. (For a description of esoteric physics, I refer you to my article in *The Beacon* of January/February 1999, Lucis Publishing Co., New York.).

MONSTERS

Has a monster or some fearful creature or person ever chased you in a dream, and just before you had to face the creature, you woke up, maybe wakened by your bed partner because you were groaning or screaming? What is that monster, and why do so many people have the same type of dream?

First, what could the monster possibly represent? Dreams are interior states, and they represent interior conditions. We can safely assume that we are working with the element of fear in this type of dream.

We already know (see chapter on Personality) that fear strands us on the personality spectrum, and we respond to fear by resisting change and trying to maintain the status quo. Most of us slog along resisting change until we become embroiled in a crisis, usually precipitated by causes seemingly out of our control. Then we feel forced into change. In a crisis we might do things that we never did before.

If the monster represents fear, then what we fear must link to change of some sort. That change must then relate to aspects of personality. We can equate the monster to the extreme aspects of the personality—like monsters of the id, a phrase used in a science fiction movie of long ago ("Forbidden Planet"). Roughly speaking, Id was a term used by Freud (roughly) to characterize our animal instincts that influence our behavior.

Monsters affect our behavior. We can see them residing

at the extremes of any personality spectrum, sitting there, ready to go into action. (Please see the listing of personality characteristics in Personality (Figure 24) and note the characteristics at each end of each spectrum: monsters, indeed.). The monsters make the spectra. In our more primitive states, we *are* the monsters, and during times of crisis, we might rise to the occasion with acts of great worth, or we could resort to our animal ends and act accordingly.

Do we really know how we might act in a crisis? If it were a life and death decision, and we had to sacrifice ourselves for a cause or for a group, would we really do it? Or would we resort to subterfuge, cowardice, and cruelty?

A monster sits at the ends of each personality spectra, much like gargoyles perched on the ends of cornices of a cathedral. Between the gargoyles, the spire of the cathedral rises, intuitively symbolic of the interlude of the spectra, the middle road, the way to the Soul. We all find ourselves living somewhere in this 'between' world, between monsters and the Divine.

Every day we live with these monsters. All the 'good' behavior we do, we do in the face of the monster on a spectrum of characteristics. We could just as easily not do good, and do 'evil,' instead. Nothing stops us, except for the constraints of culture or our own conscience. We do not have to obey any of those rules in our dreams.

In dreams we do not have to act in accordance with the rules of culture or the restraints of time. We can be ourselves, however that might be. Our dreams represent a truer formulation of our personal selves, truer than our daily lives. We are, indeed, chased, or haunted by monsters. The monster chases us, and we run away in fear. The monster gets closer and we respond with terror. Fear keeps us in line, and terror only means we are reaching the end of our rope, the end of the spectrum of personality. If we are caught, then we must confront this monster. What then?

We usually waken. We don't confront the monster. Children frequently have these dreams. These dreams occur during times of change, or impending change, and there is lots of change in childhood. Inherent in all our lives *is* change. If we refuse to change, to go along, or refuse to recognize the need to change, then a crisis looms upon us and forces us to accommodate. The presence of a monster in a dream warns us of change, and as such, the dream-monster serves as a relatively harmless crisis within ourselves: the warning of change that we must heed or suffer the consequences if we do not.

Children must adapt to frequent change simply from the maturation process, not to speak of whatever might be happening in their family life. When, as adults, our personality characteristics get in the way of change, then we might dream of a monster. The key is to confront the monster, to face our terror. The fear of the monster is the fear of change, at its extreme, but the extremes make and create the system in which we find ourselves. Therefore, in a big or little way, we must confront that monster. How do we do that?

With children it's a bit easier than with adults. Children are more open to doing things that they have not done before. Adults have had experience, and they already find themselves in ruts. We can reassure a child that the monster only means normal change. The child can be given the assurance that she/he can stand up to the monster, maybe by the use of a magic wand in the dream, or maybe by telling the monster to do something totally out of character, such as a dance. Whatever gimmicks we might suggest, the important thing to the child is that she/he *can* master the monster, just as magically as the monster seems to dominate the child.

We adults have grown up exposed to our cultural myths, all stated well in our textbooks in school. We have things to fear. Life is presented pretty much during the formative years in black and white terms. The good and the bad. Follow the rules or get in trouble. This makes up personality spectra stuff, but without it, we would have chaos, considering the diversity of the evolution of large populations. We find rules necessary, but they keep us on the personality spectra.

We can jump off the personality spectrum if we can break away from the rules and become a person that *naturally* lives by the rules. We can *be* the rules, rather than just obey the rules. Only then can we rid ourselves of the monster that affects our behavior through fear.

Let us say that we have a dream in which the monster chases us. What do we do next? We could analyze the dream and probably identify whatever personality quality we are dealing with. But is that necessary? Not everyone can analyze dreams, and is it necessary to identify the particular quality? Probably not.

We can look at what might be going on in our lives at that particular moment and ask ourselves what causes us the most concern right now, what upsets us. Perhaps that might help. Maybe we don't even have to do that. We might only need to know that a monster is showing its claws, and we now know that we are fearing something in preparation for change. Is that enough for you?

Don't Feed the Monster

What do we do then? How do we kill off the monster? Do we feel we have to fight it? What if we lose? Will we die? So many fears.

We can't run away, and we fear that we can't fight the monster and win. After all, that monster represents the animal in us. How can we fight something with such strength, claws, or guns, or knives? As adults, we are too far gone to use the magical tools that a child might. What can we do?

We have another way, a nonviolent way. Simply do not

feed the monster. Make the monster die of starvation. The monster will not have the energy to chase. The monster will die, not exactly a natural death, but a death at our own hands. We will stop feeding the monster.

The monster requires energy to exist, and the monster receives its energy from food. Food, itself, is energy in a form that can be consumed. (The chapter, Transmutation, discussed food.) We feed the monster when we give it energy. What kind of energy does the monster like to eat? Since it is a product of fear, we can assume that the energy involved in fear serves up as a nice meal for the creature.

Might the creature eat anything other than fear? Perhaps fear only represents its steak and potatoes. What about its broccoli and string beans? Fear comes in little forms, too. We might describe the little fears in life, the trepidations, the irritations, covering our little fears. Yes, the monster eats our anger and our irritation. The greater our emotional incontinence, the more the monster feasts. We return to the old standards of fear and anger keeping us on the personality spectra as we fear any change.

We can ask ourselves a couple of meaningful questions, and if we have the ability to look at our own selves, just a bit, then we can find keys to the larder, and we can lock the larder and no longer feed the monster. We can lock the door behind which evil dwells.

What kind of questions? For an opener, we can ask ourselves what has irritated us today? We would usually respond that some person, by his or her unacceptable behavior irritated us today. The problem resides in the other person, of course. True, that other person does have a problem, but the fact that we were irritated, points to the more troubling fact that we must have a problem in tandem with *that* person. Why else would we feel the irritation? That's a hard question to ask, because our irritation always seems so logical—especially when we are dealing with our children or our spouse. They are behaving poorly. Why should I not become irritated?

The fact remains, when there is a problem that causes friction, the problem resides in both people 50:50. Regardless of the type of problem, we will have to apply our will to solve the issue. We become irritated because we lose control. We will have to control ourselves. That's the first matter of business, to control the personality and not allow reactions to occur unless we *will* them. We must have the controls.

Does that mean we become less spontaneous, less alive? Not in any respect. We all respond very quickly, say in conversation, to a variety of inputs. Only when we begin to delve into the so-called negative emotions do we begin to feed the monster. More than that, we feed the monster whenever we do intentional harm. We could, if we wanted, distill down all the negative emotions and say that we have those emotions because of fear. When we have those feelings, we wish to cause harm. We want to hurt. We want to put fear in the other person. We want the other person to feel sorry. Fear protects our status quo; then anger appears when our status quo is disturbed; then we express the anger to hurt in some way.

A posture of harmlessness would require a dampening of all negative emotions. Then it's *not* okay to express anger, and we must always go around examining ourselves to be sure that we are not firing out negative emotions and harming anyone—even the person that we feel deserves it? Sounds sort of dull. Let's look at this issue closer.

Meditation, as an exercise, serves as a microcosm of what we may endeavor to do in our daily lives. We know in meditation we must still the lower forces of our personality and allow higher input from the finer energies. These energies that express themselves in the media of harmlessness can only express when we do not allow the powerful interfering negative feelings. In the first step we must control negative emotions. So, it's true, we have to watch ourselves a bit. We need awareness such that we control the outflow. We cannot 'lose it.'

In the next step, we allow the finer energies to express. This would take practice, until finally, we would naturally express the finer energies; and that which *wills* could no longer express the animal energies. We would have climbed off the personality spectra. The monster would die of love at the interlude of the spectrum, the only energy it cannot eat.

We engender a feeling of freedom, at ease. We no longer need look over the shoulder to see a monster coming. The only problem is that we become more sensitive. We become accustomed to nonviolence. Then violence becomes more upsetting. We cannot indulge the incontinence of violent emotion, as it hurts the body. The time for recovery from these upsets also increases.

Facing the Monster with a Child

Let us try to work out an everyday example with these principles, and let us use a child, as it all begins there. Suppose you have a child, and she comes home from school (the 3rd grade) and immediately begins to pick fights with siblings and you; and ultimately, the child has turned the household, emotionally, up-side-down. The child is screaming, swearing, calling you everything vile imaginable.

You, in the meantime, feel wronged. What have I done? What happened to the child at school to precipitate this terrible outburst of anger? Why am I being blamed? Why is this child so angry?

You make a decision. This is enough. I cannot take this disrespect. The child must get under control.

You say, "One more word, and you are on restriction!"

"I don't care about your restriction. You can't tell me

what to do. You're stupid," says the child. Sound familiar?

"Okay, you're on restriction for a week."

"So, what."

A younger sibling walks by, and the angry child takes a swipe at him, scratching the side of his face.

"Get up to your room this instant," You yell. By now your blood pressure is just beginning to top out.

The child won't budge. You shove the child, and the child calls you some profanity.

"You get up to your room this instant, or I am going to swat your bottom," you say, trying to control your tone.

"That won't do any good," says the child defiantly.

You threaten the child with a swat, with your hand raised, and the child begins to move. You chase the child upstairs as you might drive a cow with a swinging stick.

Finally, the child is in the room, and you go back downstairs. You are only down there a few minutes and you hear this irritating stomping coming from her room. The child is stomping against her door and will not stop, disturbing the entire household.

You dash back upstairs, and the child stops momentarily.

"Stop making that noise," you say.

"You can't make me. Besides, you're stupid. Shut up."

You raise your hand again, as if you are going to slap the child's bottom, and the child looks at you in defiance, but the child is quiet.

You go back downstairs and the next thing you hear is "Dad is a jerk. Dad is a stupid head." The child is chanting these phrases over and over again.

You feel your blood boil. What is the next move?

You have gotten into this thing quite far. Tempers raging. You could dash upstairs and give the child a good swat on the bottom. What else can you do? Call 911?

You and the child face much the same problem that we see in larger groups, such as countries when groups become angry at one another. Take Viet Nam and the U.S. just prior to the Viet Nam War, both countries stood defiantly face-to-face, lined up for battle. The only thing remaining was some excuse to go to war. We handily created a cause. Same in Iraq.

You run back upstairs and the child again calls you a profanity. You put her over your knee and threaten to swat her bottom.

She yells out, "No, no, please don't swat me." You set her back up. She is still defiant. "You can't tell me what to do," she says. "Bug off."

"You are staying in your room, and you can't come downstairs until you get control of yourself," you say. You leave the room again to go back downstairs.

Just as you are leaving, the child says under her breath, "Jerk."

This angry child sits between the feelings of wanting to be her own person and the need also to have the domination of the parent. There is no other explanation for the continuance of her behavior. She needs you and she, at the moment, hates you for it as if it is your fault for being her parent (actually, it is). The child has these feelings, and she also feels guilty for having them. In a more reasonable frame of mind she might admit that you have done nothing in particular to deserve this outrage.

The child finds herself caught in her own reaction, and though she wants you to leave, she also believes she deserves punishment. Does that mean that you will be doing her a favor by returning to her room and swatting her on the bottom?

If you wanted to bring the whole affair to a crisis and finally push it to resolution of some sort, you might consider giving her a swat on the bottom, not really hard enough to hurt her, but enough for her to feel that she is hurt by it—hurt feelings most of all. Many people would regard that as *not* the correct route.

You could ignore her. But she will find a way to get your attention, even if it comes to destroying things. She *will* get your attention. Ignoring behavior postpones the problem.

You can try to sit down with her and listen to what she has to say. This will work when she is not so emotional, maybe. She actually does not have anything to say, other than to say that she's angry at you, but she cannot know why. The anger comes with change.

The fears of change, the resistance to growing up, the resistance to the responsibility to grow up, all present themselves in defiant terms. "No, I will not empty the garbage!" "You can't tell me what to do!" All parents have heard these retorts. When you attempt to force a child to adhere to responsibility, then you must prepare yourself for the wrath.

All of this is part of growing up. The child's wrath is natural, but you, as a parent, want to help rather than hinder. So, you say to the child, rather than threaten her with a swat, "You must be afraid of growing up. I can understand that."

"Maybe I am," she shouts. "Get out of here and leave me alone, you jerk."

With that, you take your leave.

There, you have identified the fear in the system, the fear that gives the monster the fuel to thrive. You do not want to transfer the fear of the monster to the fear of you, the fear of you swatting her. Nor do you want her to think that she must be bad inside to have these feelings. With each of those cases you are still dealing with fear. We want to stop using fear. We know that the monster eats fear and guilt. Therefore, we must concentrate on the positive.

What is a positive aspect that we could, as a parent, reinforce within the child that might help her stop feeding the monster?

Explaining the fear of growing up, and spelling out to the child its relationship to her resistance in accepting responsibility and her resistance to authority figures, such as yourself. That's a good beginning, if you can pull it off. Otherwise, the child is liable to figure that there is something the matter with her, and she will lose confidence in herself. The skills of parenting get pushed to the limit in trying to explain these matters to a child in an acceptable manner and also being able to say these things at a time that the child has the willingness to listen. Not to mention that what we, as adults, may see as the solution to the problem may still be beyond the maturation ability of the child to realize, let alone accomplish. When that's the case, the child puts fingers in her ears and refuses to listen.

One thing is for certain, when the child is ready to listen, or asks a question of significance, that will be the time that you feel that you do *not* have the time. The child, seemingly in innocence, throws the gauntlet on the ground and says, "How important are my problems to you, anyway?" We do well to put everything else aside at that instant and sit down and talk, if at all possible. Usually that opportunity arises when we are passing by her in the doorway on our way to work or to an appointment. There's the gauntlet: How important am I to you, anyway?

When she does talk, it is likely to be in the context of her friends. Usually it involves snubs of some sort, the cruel little games that children play on each other, the same games that we see in us sophisticated adults. Troubles with friends involve self-esteem issues, and it can extend to racial issues, or just about anything. The parent can try to add the positive to the child's life.

Adding positivity to a child's life is the main job of parenting, and first and foremost it involves setting an example, the example of good adaptation shown by both parents to their own relationship. That itself sets a proper table upon which the growth of children can occur, and such a table may seem enough to good parents. Some parents say that they set a good example (and they probably do), and nothing more is needed. Is that true? Probably not, but I do not really know.

Children need tending, just like a garden. Weeds start to grow, and sometimes you have to root them out. This involves limit setting, a process necessary, but somewhat different than the need to add positive aspects for growth. What seems like something *positive* to you—fertilizer for the garden—may also seem like fertilizer from you, the male cow. She is apt to tell you that.

Limit setting is a persistent task. In large families it seems that the children take turns with their needs. Each child in turn will require and demand attention in his or her own special way. A child will accomplish this end by overstepping well established limits. Sometimes it takes the concerted effort and force of both of the parents at once, both bearing down, both laying it to the child, before the child realizes that she/he is outgunned and backs off. Pity and honor the single-parent, if she/he takes the job seriously.

We are back to the angry daughter, and she is asking you a question about her friends, and you know that she has given you a break.

"Daddy, today Lisa was mean to me," she says with a tear in her eye.

So now is the time to explain the personality spectra structure of personality to her and how fear, the monster within, brings out her anger? Get serious. Before anyone could finish, she would be out the door. She wants adult feedback, and she may not even want that. Mostly she wants someone to listen to her problem.

I want to do something positive, and I am asking myself, what can I do?

You can't think of anything profound to say; so, you decide to listen. That's the most positive thing you can do. To listen is to have compassion. From you, the child will absorb that compassion. That certainly is positive. You might even have the opportunity to slip a few hints in here or there about specific items when they come up.

There, you've done your job. Don't expect big results and grand transformations. Sometimes, as parents, we feel that we are dropping rocks into a well, and we don't hear them splash at the bottom—one of the bitter fruits of parenting. Years later you learn, maybe, what you did was valuable to the child.

If you look at the child as a receptacle in which you as a parent have some responsibility for dumping experiences, then you want those experiences, if at all possible, to be positive, and not negative, that is, not filled with guilt or fear. If you do that, you are doing your job. With each modicum of positivity that the child might absorb, that allows the child to move that much closer to the middle of the personality spectrum and away from the monsters at each end.

When faced with change, the monsters will swipe some food, enough food usually, to make noise. We hopefully learn, as adults, to move away from these monsters and grow along with the change and come out of old personality garments much as a lobster sheds its old shell. Sometimes I do feel like an old lobster.

Wisdom presents itself as a result of the Will-to-Good, and that only means that we apply our will to overcome the fears of our personality and seek out the good. We find love as the result. Will and love react, and wisdom results. Unfortunately, all of that seems a bit theoretical to the harried parent faced with so many battles at so many fronts. If you are a parent, you have a most challenging job. I wish you success and fun in the process.

PURPOSE

Within ourselves we can imagine, or sense in some way, the power we call *Will*; but when we speak of purpose, we begin to falter, especially when we try to speak of the purpose of creation. We ask, "Why?" Rarely, do we say, "Because."

Will appears more complicated than just the impetus to go to the grocery store, or our willful aspirations to save the world. If we ascribe to the notion that we can find cycles in all aspects of life, then we would have to say that Will, however we may wish to define it, plays out in cycles. The question is whether or not Purpose, likewise, exposes itself in cycles. How can we conceptualize purpose?

Purpose is the energy that sets any system into motion with a planned end result. Will is the energy within the system that we can identify, either in ourselves, or as energy in physical systems. Purpose defines the beginning and the end. Everything between we can define in energy terms or Will terms. In order to conceptualize these factors, we first need to conceptualize any system, and we will apply what we find to the universe as a whole. That doesn't sound very likely, but let us try.

System Processes

If we only define the beginning and the end, then we have linear process between beginning and end, Figure (37).

If the process is the growth of a flower. We would begin as a seed, and (in Maine) end as a pitiful frozen plant in December. Life becomes very simple if we look at it in these terms, but we know that much went into

Figure (37) Linear Process

getting from A to B. How could we describe that?

We must use analogy. We know that all energy, no matter what form or applied as whatever force, that energy exerts its power through cycles, waves, or vibrations. We could then infer that since energy manifests in cycles, that the effects of that energy, the effects of the causes, must also present as cycles. We all know the cycles in nature. Then does that mean that we get from A to B through a regular frequency wave, Figure (38)?

Figure (38) Cyclic Process

We like the neatness of the wavelike motion, and we know that biological systems can work sort of in that way. We know from experience that in biological systems what anything is at the bottom of any curve will not be the same as when the bottom of the next curve comes along. There is always change in biological systems. We have to figure out some way to put change in the system that will allow cycles but not have the restriction of the regularity of simple vibration, as in electricity. Systems change and evolve over time, and they never go back to the exact way they were before. We need to add the element of evolutionary change in the system and still preserve the character of cycles. We have really only one more choice: *spirals*, Figure (39).

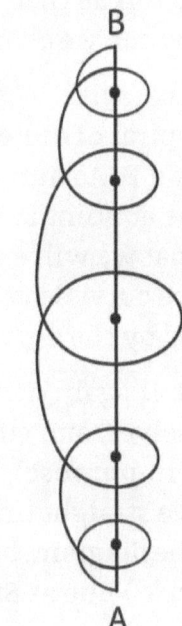

Figure (39) a Spiral

Spiral Process: Purpose and Will

In the conceptual framework of a spiral, we preserve the integrity of Purpose. The line directly goes from A to B, and for the moment we will call that Purpose. We preserve the cycles, but each turn to the cycle allows an evolutionary change. Purpose itself does not behold to cycles. Purpose is direct. The course on the spiral, however, beholds to cycles. This is the realm of Will. Will operates on cycles. If we describe Will as (c) as in Einstein's equation (please see the Chapter on Will and Space), then I will describe Purpose as (c^2), or Hyper-c (c^2), and it does not cycle. Hyper-c is direct. Hyper-c is Purpose. We live, however, in the realm of (c), which is Will.

Let us look at the spiral more closely. We live on the spiral, but purpose drives the system. Purpose provides the driving energy, but that energy becomes differentiated, without loss of energy, into forces that result in the spiral. What are those forces and how can we characterize them?

Let us look at the spiral, Figure (40) that might depict the evolutionary spiral of our earth system. We note the beginning A, and the end B. As humans, we live on the spiral. Let us say that we reside at point D. If we could predict the future, we would know that we will eventually make it to point H, and our path would be curved path on the spiral and not the straight one, as depicted by the dotted line DH.

Looking at point D again, we might say that we have a goal in mind for ourselves, and since goals aways involve a straight line (inherent in purpose), then we might create an imaginary point C, a nice straight line from D to C. (Note: That line is a little short in the diagram, but you can see that it goes off in space, off the track.) But as Steinbeck once noted, the plans of mice

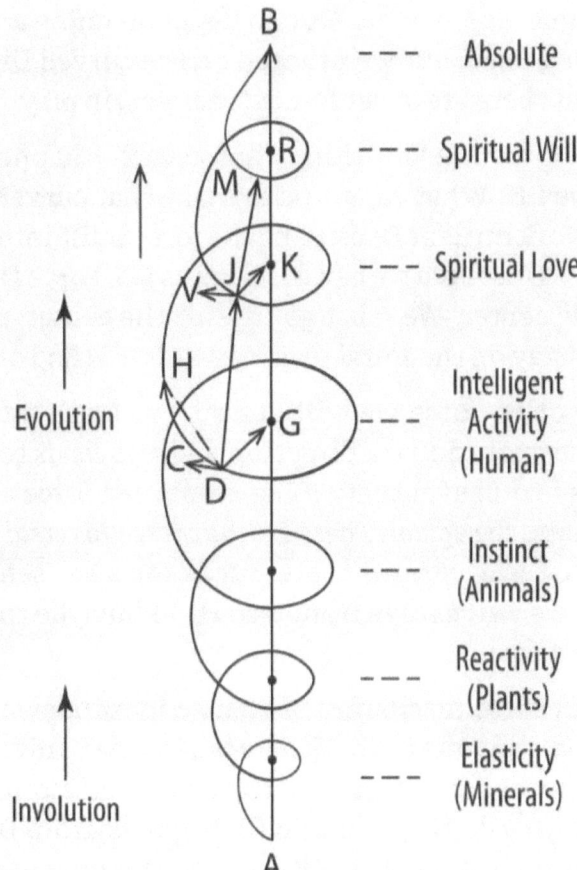

Figure (40) Evolution on Earth

and men usually don't work out so well; and we won't eventually get to C. We will go to H. We cannot see H ahead of us. We can define a force, describing the line DC, that is our little will (our goal) trying to reach the imaginary target, C. Unfortunately, we won't arrive there, but we will attain a modification of that, which we call H. We can term the force of vector DC, Desire.

We might *aspire* to H, rather than C (Desire), but even if H is our goal, our path will not course on the direct and broken line (as shown) but on the curved line to H, all unpredictable and unforeseen by us. Our goal and planning are straight ahead, using logic, but we proceed on the curved line (unpredictable) as there are other forces (energies) in play.

The force of evolution, what actually happens, is the curved line DH. What causes us to make that curve? A couple of forces work on us at D resisting the force built into us called Desire (DC). One force we can describe as DG. Force DG pulls us toward the center. We must go toward the center, or else we will never stay on the spiral to go onward to B (End of the line).

The other force we will describe as DJ. DJ pulls us forward in the evolutionary direction and allows us to proceed, on our way to B, eventually. The combined forces of DC, DJ, and DG, make the spiral. *These are the three universal forces, described by all religions.* All these forces together send us on a spiral and we will evolve from D to H, and maybe to B, by the way of the spiral.

With these mechanics of change in mind, we can then ask an important question. If there is a Divine (Universal) Purpose (AB), what advantage is it, to us, if we go along with the flow and try to do things that will propel us from D to H and not keep pounding toward C? What if we put our personal purposes in line with Divine Purpose? What will happen? That is the basic, why? of existence, and if our efforts prove effective, perhaps the end result will come more quickly. We should approach B ahead of the pack and then serve as an aid to others through service. What if we refuse to do that? Ah, there's the rub.

If we refuse to go along, we will pay the price. We define that price as Karma. Hurtful Karma happens to us as we try to reach C, when we should be doing whatever is necessary to go to H. Eventually, we will learn that we should act in accordance with the Purpose of existence, because if we do not, things will happen to us, but we will work our miserable way to H in any case. Is the purpose of life to relieve suffering?

We have the choice to suffer, or to suffer. If we proceed in line with the purpose of existence we will still suffer, because our path ahead is unknown (curved), and we only see ahead in straight lines. Let us look closer at these forces that cause the spiraling.

Spiral Process and Evolution

Looking back at the figure of the spiral, we can see that progress from point A to B represents progress. How's that for a statement? But I am saying what I mean. We begin on line with Purpose and we end on line with Purpose. The spiral widens in the middle and narrows toward both ends. The middle of the entire spiral represents the beginning of human intervention, through intelligent activity, in the evolution of substance. In esoteric terms, before the middle, that is called *involution*. After the middle, the energy of Purpose (AB) we call that *evolution*. Before the middle, the felt force of Purpose is represented by vector DG, actually, from any point on the spiral with a vector directly to the line AB. Spiral throws things out centripetally, and we can see how people stray from the spiral with vector DC.

The energy of the spiral is actually the same as Soul energy, and it is the energy accessed at the interlude. It is energy from the interludes in life that take us closer to the soul. The soul merely spirals around Purpose. And we spiral around the Soul (not shown on the Figure 43).

In the first half of the spiral, called Involution, the energy affecting the spiral from Purpose becomes less and less as the spiral widens. After the mid-point, the Soul energy is spiraling closer and closer to Purpose, and the energy of Purpose is more greatly realized. That is called Evolution, and that involves humans.

I conclude that *humans are integral with the Purpose of human existence.* That means that there are humans, and there is also human Purpose. Remember now, I am only talking about energies, and right now the big one: Purpose. I use that exact word, as there is no other word that I can use. The problem is that the materialist thinks that if you have purpose, then you have to have a personage, or a God, or a giant Salamander that has that Purpose. Yes, you can have that, but you do not have to have that. It's either way, and we are getting to the answer to how it can be that way.

Each turn of the spiral represents a new "level" in consciousness. Comparing point D to point J, we have the same relative position, except that J is closer to the center line and also closer to the end, B. We can name these spiral levels, as we can see that matter (or substance) evolves to become more reactive (more conscious) as we approach B.

In the first turn of the spiral, substance develops elasticity, the property that we can ascribe to the mineral world. In the second turn we see plants. Plants can react, in their slow way. On the third spiral, substance has evolved to the property of instinct. We see this in the animal world.

As substance develops the characteristic of instinct in the animal world, we begin to see the greatest advance of all, that of intelligent activity in humans. Beyond the human realm we see the two layers higher, first the so-called Spiritual Love level, or the Kingdom, characterized by the quality of Spiritual Love. Following that, we see the probable final level of human evolution, characterized by Spiritual Will. At that point, we are close into attaining One with Divine intention or human ultimacy.

We still have the assumption that the purpose of existence, as we know it, is to evolve from point A to B, and the purpose of any esoteric or spiritual training (or religion) is to speed the process. How can that process be speeded? We could go faster, that is, proceed from point D to any point on the spiral, such as point H, faster. That would mean that we choose actions consistent with the forces of evolution. To do so would require that we forget our personal force of Desire (DC) and bet on aspiration (to H). That is a good way, it would seem.

We might consider another way if we note that on the closest point of approach of the spiral beyond us, that of Spiritual Love, we can place the hypothetical point J. Point J can have influence upon us. Point J represents the closest point of approach of anyone of us to the next major level of consciousness beyond us. We could, perhaps, touch or access some of the energy available at point J. If we saw, even for a moment what can be seen at point J, then we could make our course up the spiral even faster. This is called, in esoteric terms, *initiation* or some call it *inspiration*. Through the attractive force of the next level above us, humans can realize that level. Such an experience sets the tone and direction for the next period of that person's life and speeds evolution for that person, but it also speeds karma, and may even result in an early death.

We can characterize the three principle force vectors (DG, DJ, DH) in terms of Good Will, Will-to-Good, and Will of God. We can say that Good Will represents actions congruent with the curved vector of DH. Will-to-Good represents the vector DJ, that which pulls us forward on the evolutionary march. Will of God (or Purpose) pulls us in toward the center of rotation of each spiral turn, the closest vector and highest energy approach to Purpose.

We could define the centers of Purpose as shown on the diagram as G, K, and R. Each of these centers resides in line with the Purpose of creation, but if we define evolution as the trip along the spiral, then points G, K, and R, position themselves outside that process of evolution. We will discuss that process when we discuss the concept of synthesis.

Being

We must pause to draw a distinction between Will and Purpose and briefly bring in the concept of Being, lest we have confusion. If we add up all the Will forces, DJ, DH, and DG, the summation of these forces (All Will) is the same as the energy we find within Purpose. That is to say, to go from D to J on the spiral requires all the combined forces of Will, but that is the same distance (if we can use that term with Purpose) between points G and K. Though Will and Purpose are one, and compose the same energy, they are related by the exponent 2, as in c and c^2. Purpose is more direct, by the exponent 2. More direct to what?

Now we talk about *Being*. The line of Purpose connects the beginning with the end, but the beginning is One, just as is the end. That is Being. Therefore, One proceeds to One, in line with Purpose. If all is One, then why is there anything in the middle: the spiral? That is, again, the why? of existence, and that will remain a great mystery. We can suppose our purpose, on the basis of what we already know of our existence on earth, but when we consider the end result, then we draw a large blank. On the spiral of life, Purpose relates to Will in the ratio of c^2 to c, but Being is immediate. Perhaps c^2 is as immediate as immediate can be, for us, and quantum mechanics allows for the immediacy of entanglement, likely beyond speed of light.

Spiral Process and Time Dimensions

Going back to the spiral diagram, we could also define these forces in terms of time dimensions. Our course from D to H is Linear time. (Please see the essay on Time to explain the dimensions of time.) The force vector pulling us onward, more directly in concert with purpose, but still within the sphere of evolution, shown as the vector DJ represents Abstract time. The force pulling us inward toward the center, shown as vector DG, represents Spiritual or Purposive time. That which resides on the line AB represents Zero time, the Eternal Now.

We can also say that the force pulling us forward is one

of Love. The force pulling us in toward the center on the spiral is Will. The resultant force, or the result of these two forces together we could call Wisdom, represented as actions speeding us on the path DH, and not DC.

Spiral Process and the Three Universal Forces

Concerning the path DC, that of desire, we can see that it leads nowhere. It does lead to destruction or obliteration, and automatic karmic influences which will eventually force us into realizing that DH is the way to go. The Will of God vector (inward), DG, continually opposes the desire vector, DC. The force of DG will always overcome, eventually, the heart of all evil: desire. Hence, we have optimism.

We have three basic forces, as shown on the diagram: Out, represented as DC. In, represented as DG. Up, represented by DJ (I say up, only because it is up in the diagram). Such a situation holds true at all turns of the spiral. Each turn of the spiral has its own set of triadic forces. Human actions going from D to C will be pulled in line with H. Actions in the direction of V will be pulled back on to the spiral—all by the action of karma. Karma does not stop its effect until B is reached. That also means that we do not see so-called perfection at any level until B is reached. Even then, this entire system only forms part of larger systems: the solar system, the galaxy, and ultimately, the universe. At point B we may find ourselves starting out again, merely realizing a more inclusive spiral, one we had been on all along unwittingly.

Our Responsibility

I have defined the influence that the next turn of the spiral has on us by our invocation of these forces through inspiration or initiation. Such influence is represented as vector DJ. In like turn, we will have influence on the spiral below us in the animal kingdom. To the animals, we are highly evolved, and our powers seem as wondrous to the animals as the next spiral seem to us. We have aspiration to better ourselves, and the next level allows for evocation to help us. When we obtain help from a higher level or we give help to a lower level, that help results from focusing energies already present in the system.

Is it necessary that we help others, either animal or human? Is our quest for speed along the spiral of consciousness an isolated phenomenon, or is our progress dependent upon our helping others at the same time? This leads us back to the discussion of suffering.

We have seen that the three universal forces pull us back in line with evolution whether we like it or not. We call this correction the force of karma; but it really is not anything more than our going against the flow. If we go with the flow of evolution, then suffering abates somewhat. Let us go back to the spiral diagram.

Vector DG has been defined as the Will of God (or Universal Purpose), a powerful manifestation of Will on the human level of the spiral, and a power that we as humans can access in mighty willfulness. This is different from DC, or the force of desire. DG is the power of the Will of God on the human level, a power, which if not tempered with love, will produce destruction: Siva in Hindu thought.

We can see that from D we can access the power of Spiritual Love, by evoking J. J, in turn, can invoke the closest point of approach of the next turn above it, M, which embodies Spiritual Will. The force of Spiritual Will acts in the same direction as that of Spiritual Love, and they work together. The key to initiation, then, is accessing that force which we can call the Will to Love.

With that bit of knowledge, we know that we must exercise the Will to Love in our lives in order to speed our own progress and temper our own suffering; and built into that is the fact that we must help others in the process. *Our* progress depends upon *their* progress. We must lift as we are pulled.

Only by relieving suffering can we abate our own suffering. The emotional aspect of such a process we call joy. **We can then safely say that as far as humans are concerned, the purpose of life is to relieve suffering.** By so doing, we act in accordance with the universal energies, and evolution proceeds as unimpeded as possible.

We work with our minds, and energy follows thought. Ideas energize, and that energy is felt in descending power on all the levels below us. A so-called *principle* is a truth that finds congruence at the highest levels of existence, and that principle effects all the levels below.

Spiral Process and Soul

We might ask, what is the soul in all this scheme? We can readily see that as evolution proceeds, we draw closer and closer to the center of existence, the line AB, and finally point B. Part of this process is a loss of individual identity and a change toward identity with the whole, One, Absolute Being. The term soul only refers to the three basic energies of the system and how they come together to form an individual, anything individual—including that block in the chapter on Will. We can define an individual on the basis of the energies that ensoul and make up that individual. As the individual evolves, that which is ensouled becomes more inclusive, and we see ensouled groups, acting as individual.

As far as we can see, we stand at the center of the evolutionary process as shown on our little spiral. Therefore, it is critical, or rather a given, that we naturally fit into this process. Something within us points us at least in the correct direction. Granted, the universal forces will guide us (or force us), but we are part of the system, too, and not something totally foreign.

That part of us which allows individuality such that we can exist as living people is the Divine Spark of Purpose. In esoteric terms it is called the *monad,* but it serves as the nidus of individuality, and it begins a focal point that allows the intersection of energies and the subsequent collection of energies that we call the soul.

The inclusiveness of this ensouling process becomes greater and greater as evolution advances until the ensouled entity becomes something as large as a world, a solar system, but not quite a universe—that is the realm of the Ultimate, we think. We are, indeed, atoms or building blocks to larger units. When we act in accordance with the evolutionary forces, then we become true to the organism of which we make up only a part. The entire organism evolves. We can say the same of the little lives that make up our own personal organism. We represent the end or the result of this particular evolutionary stage for them. As we evolve, so do they.

Perk up, you atoms of my pancreas, you have a future, too!

SYNTHESIS

Creativity is the key to the universe, and we work from reality. The instant of now is real. Anything before that--anticipation, and anything after that--memory, neither being real. The now is real, and it is eternal.

Most of the time we operate under the influence of prejudicial thoughtforms based on past experiences or anxieties about the future. Creativity reaches beyond our time and space into the Eternal Now as exemplified by the instantaneous flash of intuition, and it takes us beyond our conjurings and misgivings concerning the past and the future. Creativity taps universal energies with Plan energy and produces truth from within ourselves.

Contemplation is the creative stage of meditation and takes off at the end of a line a thought, where we can go no further. We take a step into the unknown and access the intuitive energy that gives us an answer. All questions receive answers. We find that *the familiar* veils important reality, but we cannot realize such reality unless we have a cognitive system or systematic slot in which to put it. The intuitive flashes build our cognitive capacities so that the familiar takes on new reality.

To set the table such that we can realize what exists beyond the veil of the familiar we must position ourselves properly. We can do that if we foster three attitudes within ourselves:

1) That we all share the same human tendencies, to a greater or lesser degree,

2) That whatever we fear or despise the most also exists in ourselves, and

3) That our past attitudes and thoughtforms have little use for us now, and we need to investigate new patterns. Scratching your head on that one?

Such a positioning of ourselves raises the tension of our own systems and lends to the possibility that we might learn something new. We become more spiritually 'intense.' We must move away from our usual preoccupations, because such are 'extensions' of the past that steer us away from our goal, which means that we must step beyond the obstructions of our own personalities. We stumble over blocks protruding from our personalities.

We can become aware of our personality problems. If we can deal with what appear the greatest problem of these, only the greatest, then progress will result. Generally, that problem involves self-centeredness and illusion. Allow me to use one of my own dreams as an example.

I had a dream in which I found myself in a situation where I thought I needed to make a good impression on a king. I felt insecure. To boost my value, I put on a different shirt, a very tight shirt made of wide fish net. It looked horrible, with my flesh bulging from the net openings, but that was what I thought I needed to do. The king politely ignored me. I felt frustrated, unimportant. I thought: What else can I do? I did not know what else to do other than act an even greater fool.

The dream meant to me that I wanted to impress with something other than what I really was. Something much uglier resulted than I ever anticipated—an exaggerated ugliness of myself. From the dream I learned that I can only *be*, not act. Otherwise, I am fooling myself, and others could readily see through my folly. I could not afford the luxury of such folly if I wanted to approach being.

If we position ourselves in order to increase spiritual tension, we then find that we stand between the spiritual and the physical every moment of our lives. We must use will, like a lens, to focus the spiritual energies onto the physical plane of existence. In this process of applying the spiritual--this process called *service*--we also receive revelation. Revelation comes to us because we are transmitting (and transmuting) these energies and applying them. We will not receive these revelations outside of the creative drive to serve, and your art can be that creative drive to serve. Crises lead us to points of tension. Karma being what it is, if we do not create our own spiritual crises, then they will be foisted upon us.

The concept of service that I talk about (above) does not mean that tomorrow you must drive to the Salvation Army depot and dedicate your life to that fine organization. Nothing the matter with that, but I mean something else by service. Service means doing what you normally do, but with the added Spiritual dimension—the joy of really doing what you do with all your best inner resources. Is that understandable enough? BE what you are. If you think about it, all jobs are service jobs. All work that we perform is service to someone, and we get paid for it. Nothing the matter with getting paid for service. If you cannot BE yourself while performing your service, then it is time to move on, or try to change the system. But it takes courage. That's what it's all about.

We can call all the experiences we have had, and the feelings, fears, anticipations that we take to each new experience, our 'extensions.' We can, if we want, examine our feelings or thoughts pertaining to any new experience and trace backwards from these extensions to our center of being. All the extensions reside in time and space. Our center of being is out of time and space. From that center, we can experience the spiritual aspect in substance, the quality of Divinity or Purpose in our solar system. More on this later.

Spiritual training can help us use that so-called third eye and experience vision. A vision consists of an idea developed through creativity that combines the love-wisdom from the soul level with the Will-To-Good from the intuitive level, thereby bringing to life the idea. This process, when applied in the form of service, we call *redemption*, a process that enriches the world, changing it for the better by redeeming the physical. As vision translates into reality, the physical and the spiritual move closer together, toward at-one-ment (atonement).

Relativity, Duality, and Triplicity

Judging ourselves or setting our priorities within the realm of time and space only exposes our frailties as personalities. As personalities, we have weaknesses as well as strengths. Our being, in-itself, resides beyond our time and space parameters, and it is a *non-relative* point out of time and space. When we talk of being, then we go beyond the relativity of Einstein and the conditions of the elements of consciousness.

Relativity exists on the physical plane of existence. Upon our everyday experiences hinge our hopes, fears, and angers. In this sphere we see life all in relative terms. In the most mundane terms we compare what we have to what others have. Satisfaction never comes, because someone else always has more. Whatever emotion we might have, we can relate that to wanting something. That's desire, and desire keeps us at arms' length from our center of being.

As our personalities slowly, and through much effort, become more infused with soul energies, we then find ourselves sometimes here and sometimes there. We may reside in our consciousness anywhere between, as the soul and personality energies begin to merge, Figure (41).

The position of our state of consciousness remains relative to either soul or personality qualities. We see this process within the parameters of relativity—duality, in fact.

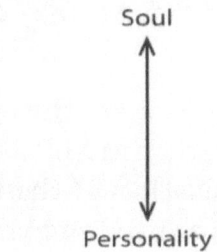

Figure (41) Duality

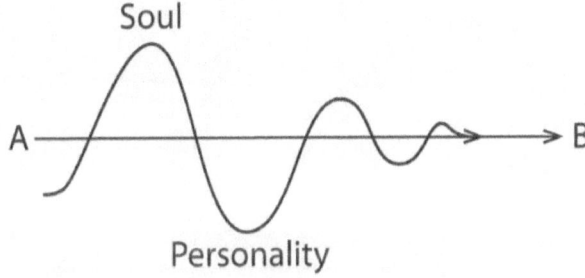

Figure (42) Unrealized Direction

The rhythm between personality and soul rectifies into an unrealized (to us) direction (B), Figure (42), combining the parameters of soul and personality into our everyday lives. That is normal everyday living.

The direction described, AB in the diagram, is a direction that we ordinarily only realize through an examination of our history, but it exists in all our actions. We might term this direction the realization of the aspect of Will. Even though we do not ordinarily realize this direction, we still reside within the bounds of relativity as we are negotiating with the elements of consciousness.

Nearly all of us have the ability to know and integrate two simultaneous factors, the example here being soul and personality. The integration of three factors is another matter, however. Try listening to two sounds, and you will find it easy to integrate them and keep them in mind. Then try three sounds. Unless you are quite unusual (as is a jazz drummer), you will become disoriented and lose all three in a jumble (a jazz drummer can do five, I think). Our lesson here regards relativity.

In esoteric terms we have the relationship of three energy combinations, which in a trinity, compose an individual, Figure (43). The organization of the three elements of consciousness contribute to an individual: love, organized by the soul; thought organized by the

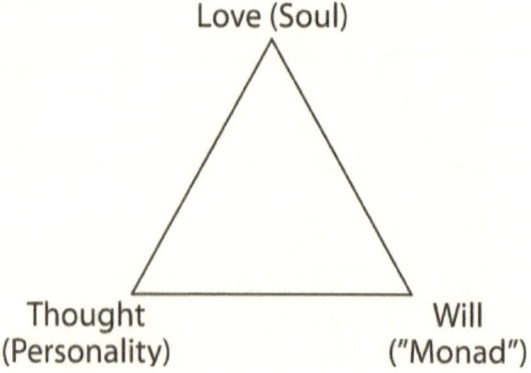

Figure (43) Trinity of an Individual

personality; and will organized by the element within us that defines unity, the so-called monad.

If we see ourselves as (human) organs that direct universal energies and create with the use of these energies, we will find that we form only a portion of a larger organism, and that organism in turn receives energy from sources outside, just as our bodies breathe air and take in food. All is energy, and we face larger and larger integrated systems as we approach the concept of the universal, representing Oneness.

Each individual entity finds the source of its being from a more inclusive (grander) entity, until finally we have a universal entity. The source of this Being is out of time and space as we know it. Within our realms of time and space we objectify our being through our personal desires. Each more inclusive entity has its own realms of responsibility, and the lower depends on the higher just as the higher depends on the lower. Only when we reach the ultimate source of all this energy do we hypothesize something that has no dependence. We can conclude that so-called perfection can exist only at that ultimate level (the *entirety* has perfection and not necessarily the individual elements), and all entities no matter how grand that exist *within* the ultimate concept of Being must have karma to work out. Maybe you could call that Original Sin.

Being and Synthesis: Beyond Relativity

Relativity does not exist in such a system as our being. The source of life is the same for all. Being permeates all Life. Everything in existence has life. Variation in forms of life we define in terms of consciousness—a rock with little consciousness, a human with, hopefully, greater consciousness. Consciousness deals with time and space. *Life deals with duration, but not time.*

Karma operates from outside our line of being. If we act according to our being, then we know that we act in accordance with the being of the larger organism in which we might find ourselves—be that the family, or country, or the universe. All being is the same. We are a cell acting properly. We do not act as a cell gone wild, like a cancer cell; and accordingly, the larger unit of which we have membership, feels better, acts better, and can act to fulfill its being. Such is redemption up and down the line.

We are agents of karma (cause and effect) to the larger organism, just as our cells and organs are agents of karma to us. We have direct responsibility for all organisms less conscious than we, and we can consciously affect those organisms in some way, to redeem them, to aid them in acting according to their being. We therefore want to take care of our bodies and all it contains, and we remain stewards of the earth and all it contains.

We might say that all levels of Being do pertain to relativity, each bearing relation to the next. On a physical or substantial level, we would be correct, but on a Being level, not so. We find One energy of Being bringing life and ensouling larger and larger entities. This Being pervades all aspects of the universe, but remains intact, even though it manifests differently at different levels of consciousness as we will see.

We, as humans, have specialization of intelligence, but unless we use this intelligence with heart, that is, we distribute energy with consciousness in accordance with our being, then we will not redeem; and we will suffer because of it, as will the larger organism of which we form a part.

The process of synthesis—our accessing being and acting accordingly (called redemption)—is a more inclusive process than relativity. We use the powers of consciousness to express life through creativity. That new life, through creativity, is what redeems. Being, and its consequent life, remain separate but integral to the processes that depend upon relativity.

Heart

Within our particular system of earth, we are dealing primarily with the energy of love. That is only to say that the road to redemption within our particular system will consist of our working primarily with the energy we call love. In other systems perhaps somewhat like ours, elsewhere in the universe, we might find a different redeeming energy. We are stuck with we call love energy, and our goal in life is to modify our given powers of intelligence with the use of love and thereby transmute intelligence into wisdom. We can then help the larger organisms (or organizations) of which we form a part.

Heart becomes most important. We have the gift of

intelligence, but we do not automatically have heart. Heart ensures the wise use of our intelligence, to circulate energy consciously in accordance with our being as we realize it. When we access Plan, then we access the love-wisdom within consciousness, the level that we immediate aspire to, and that which positively encourages the evolution of the universe.

When we have heart, we raise ourselves, but only through service (through kindness), which is the application of love-wisdom to everyday life. We may find ourselves in a point of tension between soul and personality, and that is good, but we will not find love-wisdom, the energy of Plan, unless we take a creative step beyond that point. That step, from our perspective (not the soul's), appears as a step into the dark. We will see.

Elements of Consciousness

As we know, the elements of consciousness are will, love, and thought. Will, intention, and motive, all partake of the same energy and give **direction**. Desire, feeling, imagination, aspiration, and love derive from the same energy, and **all draw things together**. Intelligence, memory, logic, and thought partake of the same energy, and they tend to **categorize things**, mostly by analogy. *Creativity, as we will see, synthesizes all three of these energies, the energies of consciousness, and exposes Being.*

We end up with the conception of so-called rays or templates, or bundles of energy of which each of us has composition. These templates move us in particular directions. We have certain abilities and potentialities. Esoteric thought calls these *rays*, to emphasize that we are trafficking in energy. At the soul level we relate with other souls of the same ray quality, but within our personalities and souls, we are each a unique mixture, as unique as are our fingerprints. Let us now discuss each of the elements of consciousness.

Will

Life does not have quality. Either life exists in an entity, or it does not. In the very basic units, life seems always to exist, such as in atoms. Atoms have movement, and as far as we know, they always have had it. We define life in terms of the ensouled unit, whether that be a human unit, or a cell, or a rock, or a universe. We can view Life as always present, everywhere, as it depends on Being. Ensouled combinations of matter pass in and out of existence, yet life remains in all the components.

Life within an organism is an ensouling life. The atoms and molecules do not die within us when we die, but that collection of atoms and molecules that composed us as living human beings is affected by withdrawal of the life force. The human dies, not the atoms. Life either exists or it withdraws, in which case it no longer inhabits or ensouls the entity previously inhabited. We associate life with movement. We see, over centuries, movement in certain directions, at least in the human realm, facilitated by the creative inventions of humankind. That direction we call will. Will gives direction.

The will we produce in our lives varies according to our own inertia and the friction or resistance we encounter. Life involves movement and is the power of the organism to move and function as that organism. Will describes what happens when that organism begins to function as a live conscious ensouled organism. Will is a manifestation of life once the organism begins to function, but it is not life itself, as life continues to exist within the aspects of the organism both before and after death.

The power of Life is the power of Beingness and is the primary power in our universe. It is indivisible and permeates all creation. Life itself exists beyond the creations, but when manifested by a creation, we see evidence of this ultimate power through movement and will.

Whatever created the universe has consciousness inclusive enough to encompass the entire universe and such Will, the Divine Will (if you want), is the Will of the universe. Our wills partake of that 'Divine' Will, and we are no more and no less than that, though we use that will according to our motivations of the moment. The power of our wills, nevertheless, is that same Divine power, and that power has the coloration of Divine consciousness (motive) even though we might pollute that motivation with our own personal desires (and pay for that through karma—cause and effect). Over time we see that the fundamental coloration of Divine Will wins out, and history slowly evolves in a direction.

When we take our step into the unknown and into creativity, we are using our wills more in accordance with Plan, and we call that the Will-to-Good. That is not exactly the Will of God (Universal level of energy), but it burns of a Divine flame more than Good Will, and much more than good intention.

Love

We have examined Will and Life in relation to creativity, now let us turn our attention to the power of Love. The energy of love colors God's (or Universal) Will. All emotion partakes of the energy of love. We turn the energy of love around to meet our small needs, and it comes out misdirected frequently; but we must accept that the power, the energy behind all emotion, is love. All the so-called negative emotions are perversions of the basic energy of love. But why emotion in the first place?

Emotion develops as consciousness develops, and the colorations of emotion have origins deep within our animal selves. We cannot just shut off tendencies from the past. We must go through the work of transmuting these old selfish feelings once quite productive and valuable for survival in our animal pasts, but now bothersome and sometimes destructive. Except for the energy itself, which we call love, much emotion is just old stuff that we must work off, the old tendencies, instincts that now present themselves as emotion. As such, love is not an emotion, but a power. Emotion is this residual animal stuff. It makes life colorful and surprising. That is about all we can say for it, but it is the mortar for stories.

Love is a power. If we want to characterize this power, then we could call it magnetic, much as the ancients did. In systems with consciousness, love brings together and accepts; and in this we see exactly what goes on in the process of creation. Disparate parts come together through motivation to produce something of worth. The power of love results from the conscious manifestation and application of the Will of God, or Universal Will, or the level of Oneness, or Entirety--all the same.

Thought

Thought comes about in relation to consciousness. We see consciousness of all grades throughout all of creation, from mere elasticity in simplest atomic systems, to instinct in animals, and self-conscious thought in humans. When any changes occur in the physical, we see an orderliness or direction for the whole, and we call that orderliness a product of thought.

We cannot assume that the energy configurations of nature, the universe, are merely random. If they were, then we would have no thought in the universe. We know that we have thought, and we (at least I) believe that analogy holds true in the universe. Therefore, to say that the universe is random is inconsistent, illogical, and most likely not true.

Thought is the product of mind. We must not get confused by the word "Mind." Some languages do not even have a word for mind. When we talk of mind, we refer to that which produces an effect of any type. We always have to return to whatever does the thinking. We call that individuality. Individuality actually has its origins with Life and Being, and not with the elements of consciousness.

Let us explore individuality just a bit.

Within any organism that has a sense of 'I-ness,' that sense of individuality extends throughout the entire organism. Previously I defined individuality as a quantum of purpose. All parts participate in such individuality. I say my liver, my hand, my cells. Each of the units making up myself works to maintain my entire organism. No part says that it wants to be something on its own. If that does happen, then we see the phenomenon called cancer, or cells gone wild, madly attempting to reproduce themselves with the result of killing the entire organism and ending the sense of I-ness ensouled in that physical body.

If there were a "spiritual" cause to cancer, it would associate with a loss or lessening of the 'I' control over the body which would translate into things that we can control ourselves: lack of physical awareness and control, depression, futility, poor nutrition, poor uses for the body—all possible contributors.

We have seen that the products of Divine consciousness, those products being thought, love, and will, each of these energies pervades creation and comes together to form greater and more inclusive organisms. We see these Divine energies in all aspects of the universe, including the energies that we use in our everyday lives. We have also seen in 'higher' organisms enough consciousness for individual responsibility, and this decision-making and energy-receiving facility we call mind. We become individuals, but I have not shown, yet, how Being produces this individuality. I will.

So, is that all there is to the universe?

Substance

We have one last component, called substance (matter), and we know that energy and substance can interchange by use of the highest energies known in the universe. Everything that exists has substance, rare or coarse, at all levels of consciousness. In relativity, as per Einstein, we see the relation of energy, light, and matter, and we know that energy can convert to matter and visa-versa.

If we hypothesize the existence of Divinity, then the universe as we see it, is the product of Divine Thought, basically Divine Consciousness, and the energy of creation and evolution follows this Thought. All of this still keeps us in the realm of *relativity*. Consciousness and matter and the interconversion of the energy of consciousness and matter concern relativity. As was said before, *synthesis, and not relativity, involves Being.*

Where did all of this substance come from? We can only hypothesize that it has existed as long as the universe has existed, and that maybe it existed even before this universe, perhaps within another universe, in which case it might even have the karmic imprint of that other universe. If that is true then we have made the analogy from humans to universes, in saying that karma (cause and effect) drives the creation of universes just as it drives individual incarnation. All pure speculation, of course.

We have discussed what makes up the universe: the elements of consciousness (love, thought, and will), individuality, substance, and a little about Being, and we want to bring in some more discussion on creativity, and a bit about the soul before we launch into our subject of synthesis.

Creativity

In creation we synthesize a channel. A channel is an ordering, by use of mind, of energy configurations that allows the flow or attraction of additional energy. We might use the analogy of the lightening rod. Lightning strikes from the cloud to the rod, and thence to the negative ground.

We humans work from the negatively polarized ground, and thence to the lightning rod (a mere channel) and thence to the positively charged cloud. We build such a channeling in our lives through the use of the elements of consciousness and through meditation and service, utilizing all the hard fought-for and innate qualities of our soul infused personality to build this channel. If the personality is not sufficiently soul-infused, then the energy imparted will destroy us more than it enlightens.

I might add here that, as I said before, I held off releasing the material in this book for 30 years or so, as I thought there needed more time before anyone would accept it, or even read it. That's true. But there was also another reason. I found at that time that the energy input was beginning to adversely affect my physical health. If I kept going in that creative vein, I figured I would destroy myself, as my own personality was not sufficiently soul-infused to withstand the energy input. I decided to withdraw and just live life and serve, as best I could, and learn from life. Well, that's what I did, and I'm 30 years older, and at least I am still alive to be doing what I am doing right now and feeling good. That's a victory, of sorts, I suppose. I did write a few books in the interim.

Back to the matter at hand: creativity.

We construct this channel for creativity, connecting with the level of Plan. Then, POW! We make a connection, and it all comes together; and we have the idea of a creation. We then have to work to express this idea at our everyday level.

In creativity, we access Plan. If we had the ability to continuously contact that area, we would have a channel made up of substance that connects all the way from the physical level to the Plan level. That channel would consist of substance from all of these levels, and it would stabilize, without the POW! more like a series of POPS. We would have the creative eye (the so-called third eye) open all the time.

Such energy input in everyday life, life being so complicated, would be a lot to handle. Psychosis can result, and in fact, does result when this type of input comes to a personality unprepared and ill-equipped. We can say that creation manifests Plan. Creative thought is synthetic. Synthesis is the new Yoga.

First Steps Toward Synthesis

Do we still want to take that step into the dark? From what point do we take that first step? Where is it, and what do we do to get there, and what must we do to take the first step? When we have explained all that, and we have taken that first step, then this chapter will end, and maybe this whole book will explode. A little Twilight Zone music, please.

The most important thing to remember is that to take the step we seek, we must separate ourselves from the concept of relativity. Considering all the tremendous effect that the idea of relativity has brought to our world, we still must step beyond that. Relativity tells about conversion of substance to energy, and how the energy (of Divine—Universal--Consciousness) is all. That's fine. It tells of the physical aspects of creation. It does not speak to Being. Being involves the process of synthesis. I know of no mathematical model for synthesis. When we have such a model, we will have interstellar space travel. I do not know where that last statement came from.

To take the step we want, we have to find a point without relation, without relativity. What is a point without relation? How can that be? Does not a point have to have relation to anything else within its sphere of existence?

Within any level of consciousness and within that level of existence we must have the ability to determine points, as substance must also exist in that space. To physically make a point, we need the intersection of two lines. Two lines intersecting make a point—so says geometry. Such an intersection we can call a real point, one that exists in time and space of that level of consciousness.

Point of Symmetry and Point of Tension

We have one other type of point that I will call a potential point. With this concept we begin to understand a "point of tension." More than anything, a point of tension is a point of symmetry. Let us see how units come together to form a symbol familiar to us, Figure (44).

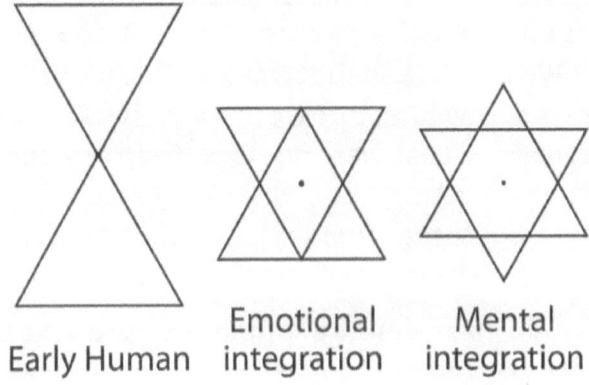

Figure (44) Development of a Point of Symmetry

In the figure of an early human, we see the symmetry basically of an animal, that of important parts, but lack of integration. The middle figure, that of Emotional Integration, represents the bulk of humanity today, struggling to integrate their emotional selves. As the triangles come together in the evolution of humanness, we finally develop a point of symmetry in the integrated human. As long as the triangles hold to that last alignment, then we have a potential point that holds symmetrical relation to all the apices (points). The point exists because of the alignment, and as such it separates itself from the realms of either triangle, yet related to them in symmetry.

It is as if we brought two lenses together and positioned them to create a focal point. Energy can focus at that point and remain independent of either triangle (lens); yet the focusing power depends on the alignment. Something is relative to something else only if we can measure an effect of each of the components of the system on each other. The alignment of these two lenses only allows energy to go through: transmission. None of the components actually makes an effect on the other during that transfer.

Motive remains of greatest importance. Without proper motive, then energy can destroy. We need awareness of our motives. If we concentrate on some of our everyday thoughts and trace their extensions back to motive, what do we find? We find mixed motivation. Yes, we are doing this or that for someone else, but we are also looking for gratitude, honor, reward. If that were not true, then any one of us would be up there adjusting our energy knobs ourselves right now. So, let us take some time and examine motivation. We are now getting to the nitty gritty of the process.

I am going to **define a "point of tension" as that point of symmetry created by the intersection of three simultaneous lines of time,** Figure (45). The lines of time can reside within one dimension of time or each in separate dimensions. **A line of time is a possible happening, a potential happening, a time passage, within any one dimension of time.**

In Figure (45), I define D as the point of tension, and points A, B, and C, as the three input energies, and lines AD, BD, and CD as the lines of time of each of the 3 different energies. Lines of time will be discussed later, but the point of importance now is the symmetrical position of the point of tension in regards to interacting systems. *Any of the following terms describes point D: point of tension, point of symmetry, point of being, zero time or the Eternal Now. I may use any of these terms to refer to that point of symmetry, depending on the context.*

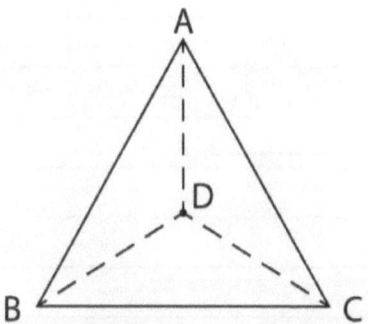

Figure (45) Point of Tension

Soul

Before we discuss symmetry, we first must clearly understand the soul, what it is, and how it functions. The soul locates itself within the higher reaches of our mental level of being, Figure (46). If we wanted to divide our mental level of being into seven interpenetrating energy levels, as is done in esoteric thought, then we would find the soul located on the highest three such levels. In general, nearly all of our conscious thought we would find on the lower four levels. (MPA and MU in the diagram will have discussion later.)

The soul is an organization of the energies found on highest three levels of the mental level of being, and the energies involved, love, thought, and will, are imputing into the soul from beyond the soul. What does this mean? It means that the energies of spiritual will, spiritual thought (pure reason), and spiritual love, before they manifest within our personalities, first organ-

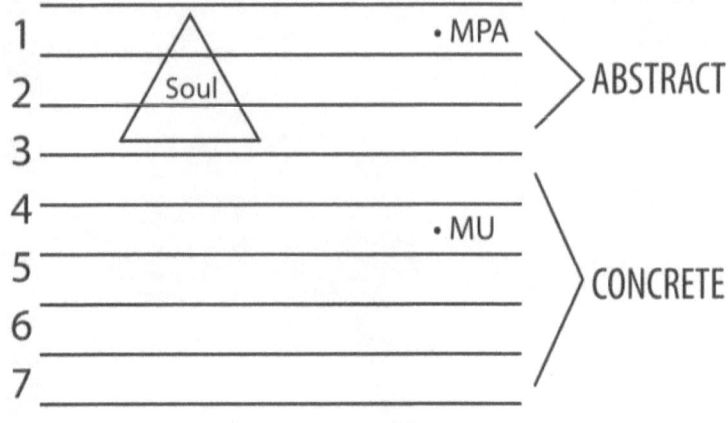

Figure (46) Soul and Mental Realm

ize within the soul. This organization forms the basis for an individual. At the soul we find the necessary organizational structure to make a unique individual. Otherwise, we would all be much the same.

Billions of individuals are needed over long periods of time to develop all the possible soul advances of the larger Soul (or energy configuration) in which we all find our Being. Each individual functions as just one possible combination of all the possible manifestations of the Universal Soul. All manifestations are necessary, and all must perfect before we see perfection of the manifestation of the higher Being in which we find ourselves. How could the organism as a whole have perfection, unless we encounter perfection of all the individual parts manifested?

A human individual is a thinking activity unit with sufficient will tempered with the direction of love, having the purpose of redemption, and seeking perfection at the physical level. Each individual portion of this universal puzzle receives its individuality from the organization of such energy at the soul level. Only when these fully individual energy patterns present themselves through soul integration of personality does the plan or potential for that individual fully work out in its capacity to redeem.

Souls not only compose indivisibly the Universal Soul, but they also associate with one another in terms of energy. Groups of souls have the same basic direction for service, the same redemptive goals. When they work together, utilizing their energies in a coordinated manner, the work goes that much faster.

We ask the question: How is it possible to have individual souls when at the same time the statement is made that all souls compose indivisibly the Universal Soul? We realize that the Universal Soul must ensoul all the many ways, in fact, all the possible ways of physical manifestation—all possible craziness, meanness, kindness—and it has divided up the job into individual human units, each of equal value and each seeking redemption on her/his own time table. After eons, the entire Unit will have redeemed itself through all its units. Is the human soul singular or is it indivisibly part of a whole? The answer is both.

To realize the redemptive goals and to realize the thought organization behind the soul, we have the potential to access Plan directly. Plan is the energy configuration of the World Soul or the blueprint of what all earthly individuals were created for—to redeem. When we access Plan, then we see the Plan beyond our personal selves. Our ability to function efficiently toward our individual (or soul group) goal has enrichment.

The individual soul contributes to the ensouling energies of larger and larger Souls. All Souls function to redeem matter (substance). The job of redemption goes on at all levels. We are merely dealing, in our human existence, with the most physical of all levels. In levels of consciousness beyond us, we would find matter of finer density, called substance. Souls group into energy centers with the same redemptive goals, but not of the same personality type.

Behind a redemptive goal at the physical level of existence we encounter individual souls. I say souls (plural), because the goal is bigger than anyone can accomplish. The goals require a group, not a huge group, but a group. The goals become more and more inclusive as the hierarchy of organization becomes more and more broad and inclusive:

What we each can do: called *Individual redemptive goals*.

Organization of a small group of souls: called an *Ashram* or *the subjective group*.

A very large group of souls: called the *New Group of World Servers*, in esoteric circles.

All groups: called *the Hierarchy of Souls—the Kingdom of Souls* (*Heaven* in Christian terminology).

We now have some grasp of the soul and how it organizes with other souls and how we cannot separate any one soul from another, as all souls have a redemptive goal. We also see specialization of small groups of individuals to accomplish particular needs.

Point of Tension and Time Dimension

What does all of this have to do with a point of tension? A point of tension is the simultaneous intersection of three lines of time from three systems. When we negotiate with simultaneous events, then we must deal with time. *Time, itself, is the relation of two events that must intersect at some point through a common experience.* At that intersection, each event would then

THE RELIGION OF PHYSICS

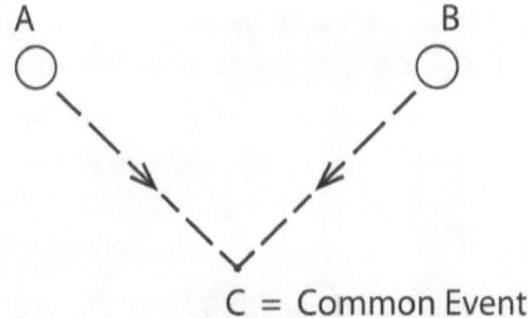

C = Common Event

Figure (47) Two Balls Collide in Space

participate in time within that particular dimension. Let us use an example, Figure (47).

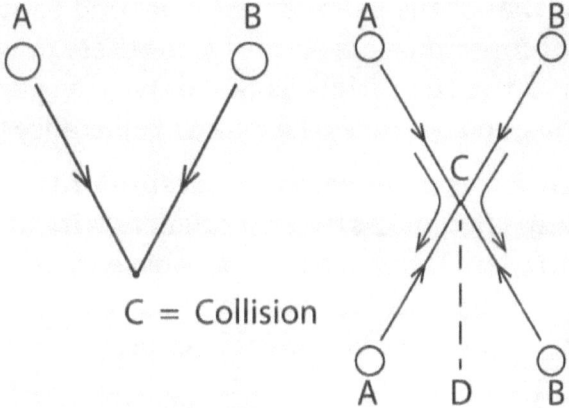

Figure (48) Balls Collide and Rebound

A and B are balls in space moving toward one another. In space, without some other object to relate to, we do not have time. What time would it be? Both balls are within the same overall dimension of space. When they collide, we have a common event. Now we have time. The time so produced, remains consistently in common with each of the balls as they go on their separate ways following the collision, Figure (48). Both A and B after a collision now have a common time. I represent that time as the dotted line CD. As long as each of these balls continues to exist, they will have a common time, but if either of the balls no longer continues to exist, then we are at the end of time for A or B, Figure (49).

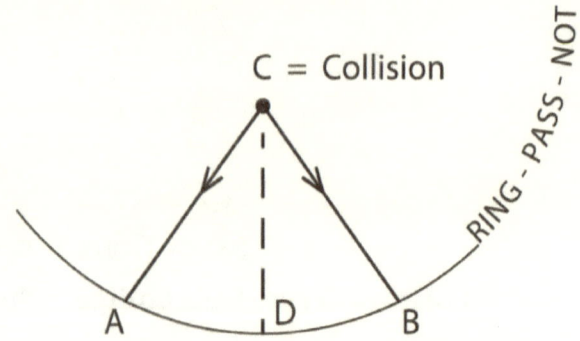

Figure (49) Out of Existence

Neither of the balls can exist beyond the reaches of its powers of consciousness (as meager as these powers are). The arc created, the end of time, is called the "ring pass not" in esoteric terms, and it defines the realm of existence for the balls.

Ball A has an experience because of the collision. So does ball B. That makes two experiences with a common line of time, CD, as in Figure (48). We now have what we can call polarity or duality. Is there another experience involved in this collision, or are the only possible experiences those experienced by either A or B individually? There is also the event of the collision but viewed or experienced by a third party. The collision viewed or conceptualized by another object is an additional event or experience. We could call that experience AB.

We have experience A

We have experience B.

We have experience AB.

THE RELIGION OF PHYSICS

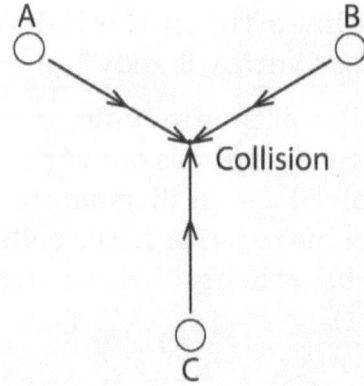

Figure (50) Three Balls Collide

Apparently, we have only two parties involved in this collision. Where is the third party, the third that can experience AB? Let us work toward that answer by examining what might happen if three balls collide in space all at once, Figure (50). Balls A, B, and C collide at the same instant in space.

We now have the following experiences:

A hitting B

B hitting C

C hitting A

A hitting C

B hitting A

C hitting B

A hitting B and C

B hitting A and C

C hitting A and B

ABC all hitting one another at once

We have a total of ten possible experiences with this simultaneous collision. Where is that which can experience the experience of ABC? To answer that we have to first determine what is the line of time in this collision. Where is the line of time in common with A, B, and C?

The only line of time in common with all the parties in the collision actually extends out of the dimension of any of the parties involved. Let us diagram this in a series of steps. First, the balls come together in the collision: We saw this in Figure (50). The balls then go their separate ways after the collision, Figure (51).

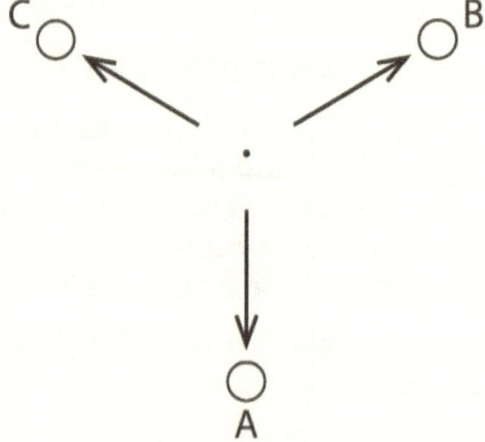

Figure (51) Three Balls Collide

THE RELIGION OF PHYSICS

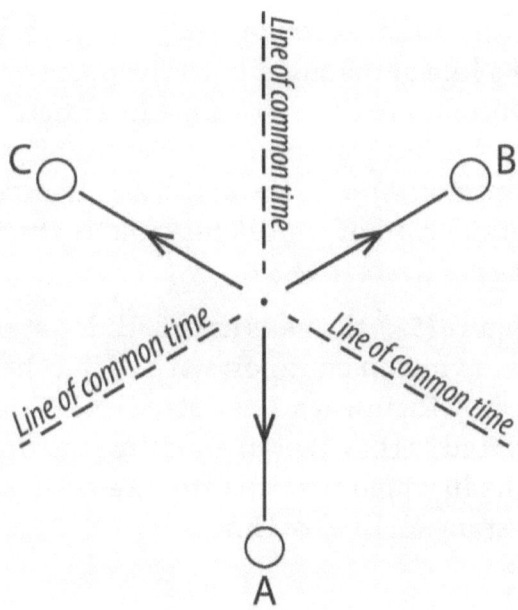

Figure (52) Three Lines of Common Time

We begin to see that time becomes more complicated in Figure (52). There exists no line of time in common in the dimension of all the balls. We see relationships between the balls, but the time in common only exists between two objects, even though we can make sense of the time by saying that a line of common time exists between, say, B and A; and there is also a line of common time between B and C; therefore, since both C and A share the same line of common time with B, then all three, A, B, and C must all exist within the same line of time. That's nice, but where is that line of time?

We were faced with this same problem when we tried to determine the whereabouts of the third party to experience AB with only the two balls colliding. Now, with three balls colliding, we are searching for the party that can possibly experience ABC (the tenth experience).

There exists another line of time in common with all

the balls, in either example, and that is graphically represented by a line perpendicular to the plane on which each of the balls interact. This line of time intersects the experience of the collision but is graphically shown at right angles to the lines of common time of the balls. We are here encountering another dimension of time. This time is the time of the instant, also called *zero time*.

In Figure (53), the balls have collided at point D and created lines of time in common with each other, and they are rebounding on their ways. The dotted line ED represents a line of time related to the other lines of time but out of the dimension of time in which the balls are interacting. At point D, for just the instant, all lines of time intersect. Zero time emerges.

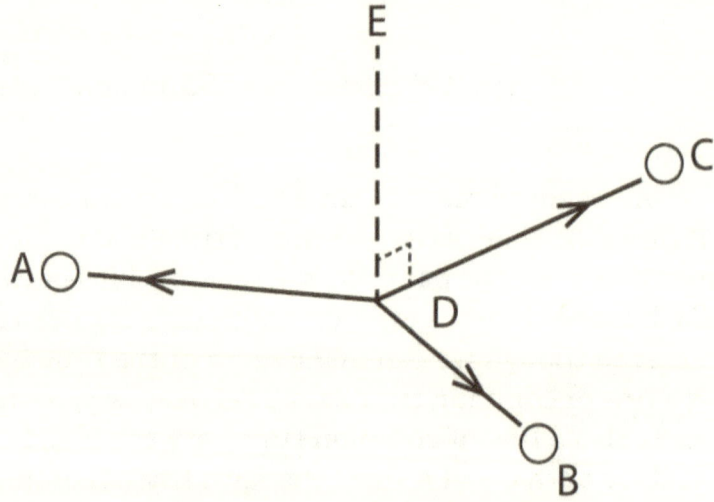

Figure (53) Zero Time Emerges

Synthetic Change

This time, zero time, has no ring-pass-not, that we can realize anyway. *Zero time is on eternal time, and it is time in common with all interactions of any kind, no matter the process.* We have synthesis only upon the line of eternal time or zero time. In the system involving balls A and B, the synthetic experience is AB. In the system with the three balls A, B, and C, the synthetic experience is ABC.

Synthesis is the momentary interaction of two or more systems. Synthesis is not the result of that interaction. *Synthesis occurs at the point of tension, which is now.* We have an infinite series of Eternal Nows, each of us.

Time is created with any interaction. We take this for granted, but if we really think about time, we realize that each event relates to something else, to something else, and to something else, until we have the final relation which is between the atom (atomic time) and the solar system. These two parameters, for the foreseeable future anyway, define our physical ring-pass-not. When we have true space travel, our ring-pass-not will expand, but for now, we are essentially out of existence beyond our solar system (though we now have extensions of ourselves—spacecraft—drifting into outer space).

At any level of consciousness, no matter how exalted, we will find time as long as we have two factors that can interact at that level. But zero time or eternal time is the ultimate interaction, with the ultimate Time Keeper (Old Father Time), and this time remains consistent, always, within all other times, and it defines the ultimate ring-pass-not of all existence.

The change as a result of any momentary interaction of now is the result of synthesis. We call such change *synthetic change*, knowing full well that something had to put all the parties of the interaction into motion in the first place. *I will use the term synthesis only to apply to the now*, which given the obvious fact that there are always more than one thing going on with us at any time (especially at our house), that every moment of now is synthetic; and the now connects us with Being. This brings to the surface the main points of all that precedes and all that follows: *When we deal with Being, we deal with synthesis and not relativity, and in all aspects of Being we con-*

tinually interact with the eternal, or the Ultimate. I will discuss a startling corollary to this last statement a bit further on.

Being and Life

We find synthesis at the Point of Being. At that point, the energy we encounter is the energy of Life. Being and Life we can consider sort of the same. *Being produces life in relation to the energies of consciousness*. We always find Being at the point of symmetry pertaining to the energy of the systems of consciousness involved.

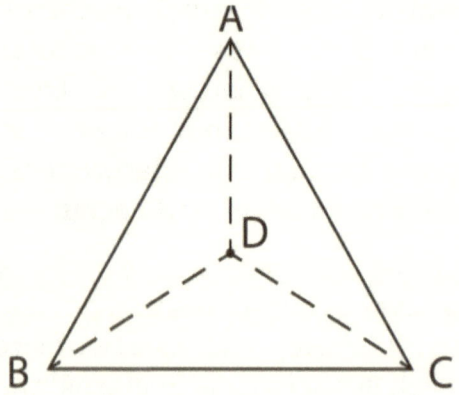

Figure (54) Point of Being

In the Figure (54) if A, B, and C, are the elements of consciousness, then D is the point of Being, and we will note that it resides on the intersection of the time lines of all three levels of the elements of consciousness.

Being will infuse the elements of consciousness to produce a Life. Existence by use of these energies of consciousness cannot extend beyond the limits of possible Being with these particular energies of consciousness. We therefore see a ring-pass-not, Figure (55), that delineates the extent of possibility of life with the use of any particular energy sets of consciousness.

As humans we have the task of integrating the energies of consciousness of 1) Personality, 2) Soul, and 3) Plan (shown in humans as love, thought, and will). True, all of these energies already exist in each one of

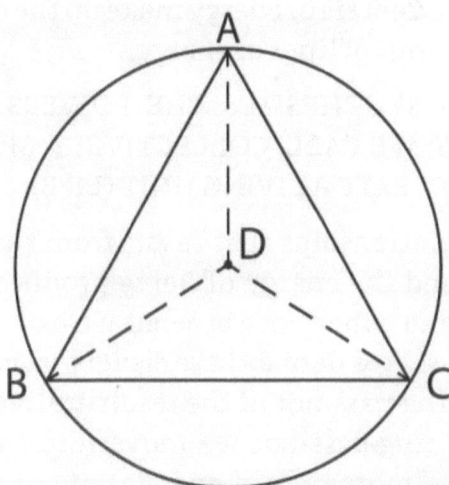

Figure (55) Ring-Pass-Not of Consciousness

us, or we would not be sitting here slobbering, or whatever we're doing. But this relation is not realized by nearly all of us. We wish to make this relationship one of conscious synthesis.

Let us, for just a moment, show the difference between Spirit and Being. We use Spirit as a general term that includes all the energies that make up consciousness. Being appears at the focus point of the elements of consciousness to produce a life. Spirit refers to the energies involved in consciousness; whereas Being is the particular energy that produces life. It's a spark.

If we want to consciously *be*, then we must consciously relate to all the elements of consciousness at once. Consciously relating three simultaneous events or energy inputs is a Divine (or universal) quality. Relating only two events or energy inputs is a human characteristic, and remains in the same dimension, just as did the 2 balls colliding. We are not surprised when we note that within the energy or etheric body of a human, the centers of energy involve rotation (within the centers). Such rotation gives us an integration of energy that we could not have otherwise through conscious assimilation. Centers of energy make up the concentrations or the intersections of lines of energy.

BEING SYNTHESIZES THE POWERS OF CONSCIOUSNESS (WHAT WE CALL, COLLECTIVELY, SPIRIT) WITH SUBSTANCE TO CREATE A LIVING UNIT (LIFE).

The relationships that result from the energies of consciousness and the energy of Being produce all the possibilities of life and experience at what we call different levels of consciousness. We define these "levels" according to the extent of the ring-pass-not of the resulting lives produced. Outside of that ring-pass-not we move into the realm of other lives that use more refined energies of consciousness. These lives we would consider "higher" levels of consciousness, and each level would have its own ring-pass-not.

The Points of Being within any particular level of consciousness are called in esoteric terms, *permanent atoms*. Permanent atoms are the centers of Being at any particular level of consciousness. A permanent atom serves as the point of symmetry for the energies of consciousness. From that point of symmetry, the Point of Being resides on the time lines of all the elements of consciousness for all of the lives on that level, and the extent (or radii) of these time lines determines the ring-pass-not for that level of consciousness.

Going back to Figure (55), ABC could be the elements of consciousness in any life system within the ring-pass-not of a level of consciousness. The energies of the elements of consciousness, A, B, and C, and the energy of Being (D), make up the energy involved in manifestation (existence or life), and it partakes of the substance available for creation at that level.

Whatever becomes manifested always involves karma —cause and effect. Karma exists because of existence, and existence occurs because the elements of consciousness come together in a way that Being synthesizes to give the gift of Life.

We must realize that consciousness in manifestation is not perfect. *Matter (or substance) carries the scars of imperfect consciousness.* Being or Life, as such, has NO taint of the karma inherent in consciousness.

From point D, all perspectives of truth are considered in the making of thoughtforms. The truth exists differently, considering the perspective, regardless of the level of consciousness. A universal idea cannot necessarily have implementation unless it first passes the group test, and finally, the individual test. Only with the finality of human processing can universal ideas have application. We see nature as a blunt instrument of broad changes. Karma induces changes, but we can look at that process merely as cause and effect. Upon the earth, only humans can apply with precision their intelligence to our physical world.

In the energy called Divinity we see Absolute Self-ness. Such an element permits creation and is the essential ingredient for creation. It is the Divine (or universal) element that has allowed humans to think separately (I-ness, individuality). Each of us is not separate. We all share Being with all existence. All existence, that which is, exists only because of Divine (Universal) self, the power behind Life, Being. It is the cause, the ultimate cause. I-ness in any unit, therefore, derives ultimately from Being. Divine Selfness is only another way of looking at Being.

We remember that I-ness or individuality is a function of Purpose. I stated previously that we share Being with all existence. We are part of nature, not separate. It is important not to become confused here. Being is an energy existing without associated substance or matter, also called energy-in-itself, and it is a Divine (universal) quality. When it associates the forces of the elements of consciousness (Spirit or Holy Spirit in Christian terms) to produce a particular life, then that manifestation of Being produces purpose within that bit of individuality —the Will-of-God (energy) built-in. This is foundational, but it can be overridden, by any of us; and as the system is set up, those reactions will not progress to fruition. They merely meet the painful obstructions of karma, the physics of life.

The boundaries and course of the system have been set by means of preprogrammed substance, preprogrammed or energized substance being the agent of karma. Space is the container for all reactions, and space is filled with substance—according to esoteric thought—and space will only allow certain types of reactions to occur: Our physical laws. Everything else fails.

We have diverted our attention a bit into esoteric physics, but I thought it important to make clear the nidus of being, our individuality, is part of Being and therefore reflects Purpose. Once we begin acting out our individuality, we can do what we want within the bounds of our physical laws, laws set by the limitations of space. The Will-of-God as a universal energy built into our individualities means that we all head in the direction of perfection, even though our courses are quite varied and some even head backwards for a while. But we all get turned around at some time through the action of karma, which is the result of preprogramming of all substance. Let us now go back, specifically, to the subject of Being.

We need to know how we can recognize Being or this element of purpose in ourselves, since this is as close to Purpose that we can get. Also, being part of nature, we all have life with Being or Purpose in common with all of creation. How can we recognize it?

The thread of consciousness, Figure (56) involves three strands or perspectives as has been said before, the individual (personality), the group (soul), and the universal (Plan). If we experience a situation, and we have the ability to simultaneously take into consideration the individual, the associated group, and the universal values involved, we would automatically center ourselves in the realm of being, the point of symmetry, the distillation of Purpose at that particular level of consciousness. What does that feel like?

The feeling is primarily one of freedom, a release from "old" thinking, and an opportunity for "new" thinking. We are clear minded in our intention and determination, and our love. We could say that these qualities associate with high levels of consciousness, but we are really saying that we break into the power of Being. Being allows life to these powers of consciousness. This stance of synthesizing the three perspectives we can also call the mechanism for creativity, intuition.

Being when added to energies of consciousness gives us Life at any level of consciousness—animal, vegetable, or mineral. Being at the human physical level, i.e., the realization of Being at the physical level, would be something quite different than the realization of Being at the emotional level, or even at the thought-level in humans.

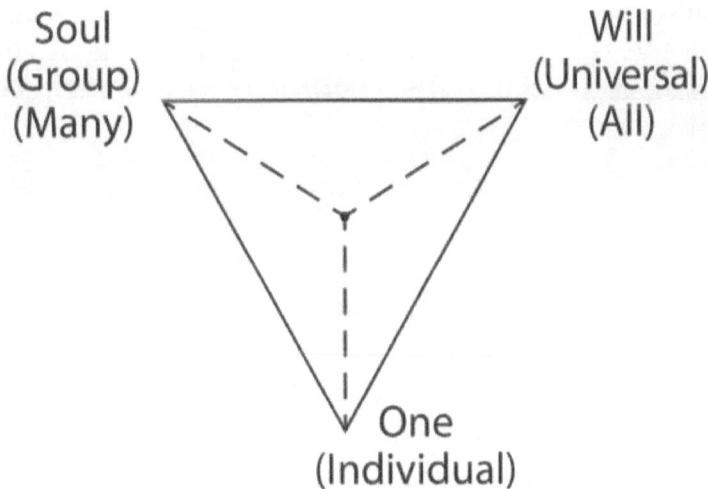

Figure (56) Thread of Consciousness

Let us try to see what Being is like at each level. Then we

might have the ability to recognize these happenings in ourselves. First, I must make it clear that Being (large case B) is *all* being, and we are only talking of how being expresses in our lives at particular levels of consciousness; and I will use being (small case) in these instances.

1. being at the *physical* level is, amazingly enough, patience. Patient, collected, poised, but not doing. We see this in meditation, also in some Olympic athletes just prior to their performance.

2. being at the *emotional* level, we might think, would consist of our "highest" emotion, whatever that may be. If we consider the matter a bit further, we see that our "highest" emotion, like all emotion, would attach to desire in the form of a thoughtform. We might say that aspiration, attempting to seek out the Divine, would qualify as being at the emotional level. Unfortunately, aspiration always taints with desire, and it remains the trap of mysticism, the desire to merge with the Divine (or the Universal).

What else might qualify for the 'highest' emotion? We could consider thanksgiving, and that might bring us closer. We give thanks, and we link to the Divine (the Ultimate); but it still attaches us to the physical, as we give thanks for something.

What is a feeling without attachment, but which allows Divine input (the input of Being), much like patience or poise sets the table, physically, for possible intuitive input? There is only one answer, and that is solemnity. Yes, being solemn. That does not mean being glum. It means the seriousness necessary for important input, not that we can't throw in a joke now and then.

3. being at the highest *mental* level is more complicated. Once we enter the mental level, we find concepts. Unfortunately, many if not most, of our concepts we color with emotion, and at the lower portions of the mental level, that

is about all we will find: thoughtforms. Emotions attached to thoughts spring immediately into our minds in any given situation. These also form the basis of the all-pervading human weakness called prejudice. We all have prejudice, all the time, nearly. That is what the ancients meant when they spoke of illusion. We can drop our prejudice only if we drop all preconceptions and face each experience with an open mind. Let us examine the mental level even further.

Ascending into the mental level, we will find a place where most of us hang our hat, and it consists of a temporary point of being within the so-called fourth mental level, called the mental unit (Figure 53 shows graphically the mental levels and the mental unit, MU.). Nearly all of us find our 'I' or sense of self (focus of the energies of the 3 levels of consciousness) at the mental unit, and the mental unit serves as the doorway to what we call our soul. At the level of being of the mental unit, we are still dealing with concepts, and these concepts we use to explain our beliefs, philosophies, and religions.

The "highest" three levels of the mental level are three really big steps. The soul also locates itself in this realm. Christian religion symbolizes these as the three steps before the Cross. We have left concepts behind at this level, and we are now dealing with the abstractions of life.

Is the statement of an abstract concept an abstraction? For example, is what you are reading now an abstraction? The answer is, No, as I will show.

When we think abstractly, we do not use the verbal aspect of the brain, as language is something quite physical, derived from physical existence. We are dealing with the mechanics behind concepts. Such is the reason that thought transference can occur between people who do not speak either's language. The transference occurs on the abstract level of thought, not the words.

Behind concepts we find structure. When we build a bridge, for example, we know that behind the concept of the bridge stands a mathematical model for the bridge. In the abstract mental levels, we are referring to symbols, but symbols much broader, more far reaching, and more inclusive than just the individual symbols of mathematics or language that break up ideas into pieces. We are dealing here with ideas with more universal application, just as a true work of art has symbolism that applies to all levels of being, endlessly analogous.

These universal ideas, from the level of Plan, filter through our subjective group (the ashram or soul group at the soul level), and thence through us, as individuals, and into concrete conceptions at the lower mental level. Once at the concrete level of thought, we can begin to put the idea to work (service). For most of us, all that goes on before the concrete level of consciousness remains largely something that we cannot remember.

We are also dealing with the soul, the realm of our three highest mental levels. Perhaps we would do well to divide up these three steps. We know that soul energy is love energy within the mental realm. This gives us hints on how the concepts involved will precipitate through service in our world. We will see them playing out as universal brotherhood, equality, and harmlessness.

If we studied these 3 steps of the mental level by using analogy, then we might say the system repeats itself, and we should approach the analysis of the steps (if analysis is possible) by saying that the first step must in some way have analogy to the physical, the second to the emotional, and the third to the mental. But remember we are still talking about thought (the mental level), and we are seeking *being at the level of thought*. I will divide the answer in three segments. Let us begin with the physical.

A. What is the *physical* analogy to an abstract thought?

This seems relatively simple. The physical aspect of an abstract thought must be the abstraction itself. That would consist of a symbol, the physical structure of the symbol.

B. What of the *emotional* aspect of an abstract thought? More than aspiration, thanksgiving, or solemnity that we found at the emotional level, we look beyond, and we find the concept of glory. I should not have used the word *concept* to refer to glory, because we cannot describe glory with words. *Glory is the feeling of the power of the Universal.* No word describes it. We must experience it.

C. Finally, the highest level, the third step, would be the highest of the mental level, and here we would find the mental permanent atom, noted as MPA in Figure (58). The mental permanent atom serves as the center of being for the mental level. Following our analogy, we would ask ourselves the question, "What is the thought of an abstract thought?"

What is behind thought? There is only one thing behind thought, and that is will. Behind abstract thought we find will, or the manifestation of the energy of Divine (Universal) Will. In humans, we call that the Will-to-Good. Hold that concept, and I will try to put this all together.

We remember that synthetic thought, or the type of thought input that occurs at the point of tension with 3 simultaneous energies intersecting, puts us in contact with our being. If we look at what is called

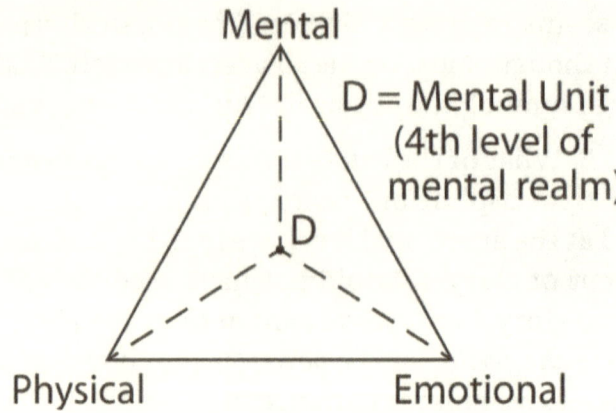

Figure (57) Being at Mental Level

the lower triangle, we see being at the mental unit level, Figure (57). We can look at the mental unit as a temporary place that we find our self-ness, yet it serves as the center of our being (the 'I') at that level of consciousness. At this level we discover ourselves and set out to coordinate our personalities in preparation to knowing the energies at the soul level. I know of no other way to depict the levels of mental capabilities of humans other than talk of these levels of consciousness, of permanent atoms, the level of Plan, and the rest. And I realize all of this is esoteric, but these concepts come from the Sanskrit, believe it or not, and are thousands of years old.

If we could conceive and hold in mind simultaneously, the physical, emotional, and the mental aspects of any situation, then we would experience being at the mental unit level.

If we look back at the three steps at the soul level that we just discussed, and we held them simultaneously in consciousness, then we would experience being at the mental permanent (MPA) atom level of consciousness as shown diagrammatically in Figure (58). We remember the components of consciousness at that level as the *symbolic* representation of the thought, the feeling of *glory*, and the *Will-to-Good*.

What do we experience at this Being level of the mental level, where we find the mental permanent atom? We call that experience intuition. Kant called it Pure Reason, the mental precipitation of what we call Plan.

We enter that level consciously only when we have fully integrated personality and soul. We cleverly combine or integrate soul and personality, making one functional unit: the soul infused personality. After that, we can then turn to the universal. So, rather than trying to hold three in consciousness, we cleverly combine two, and turn toward the third.

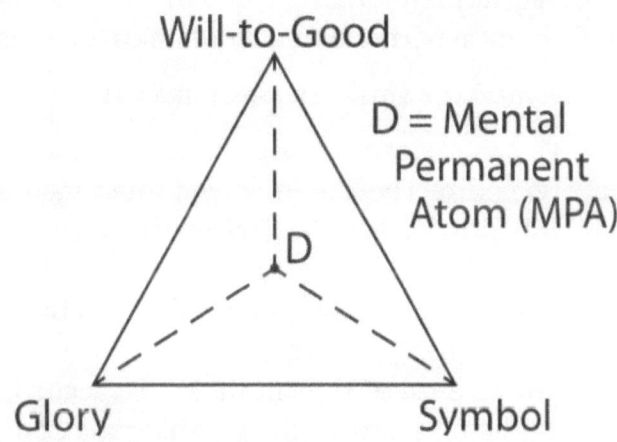

Figure (58) Being at Mental Permanent Atom

What we are seeing here concerns the elements of consciousness of two different levels of consciousness, the soul and the personality, fully integrated at long last. The point of tension created at the point of symmetry, shown in the last of three steps in Figure (46) is unique in the respect that it involves three separate dimensions of time—abstract time (the soul), linear time (of the physical), and eternal time (at the point of symmetry—Eternal Now). The result of such integration means that the individual then moves into the "ground level" of another "kingdom," starting all over once again to integrate another two systems. Beginning the next chapter.

Because of the positioning of the human soul on the upper reaches of the mental level of existence, we find no direct connection between the mental permanent atom on the highest level of the mental realm and the mental unit on the fourth highest mental level. The soul as an instrument allows intellectual growth, through love, which is the gift of individuality. Humans are allowed, in fact encouraged, and also given the means, to unveil the secrets of the universe. Without the soul to organize and channel the energies into a functioning individual, the power of the energies involved would merely blow us away. The modulation of the soul allows the unevolved human race to gradually learn. The mental unit and the mental permanent atom remain unconnected.

Beyond the Soul—the Antahkarana

We can seek to bypass the soul and access the universal energies directly. To do so, we must have a great degree of soul infusion within the personality, basically living as the soul; otherwise, the energies will destroy. Universal energies can expose and highlight the weak areas in the personality. The desire to learn how to access universal energies directly begins the science of the antahkarana, according to esoteric thought. The first part of the antahkarana connects the mental unit and the mental permanent atom.

How is the science of the antahkarana even possible, because none of us is perfect, and the energies of destruction possibly so great? As it turns out, honestly striving for the goal protects us. With a moving target, so to speak, disease and injury are not allowed to 'settle in.' Rather, our positive creative-will neutralizes those energies that would otherwise overstimulate our defects, and the healing equation continually drives forward. We speak, however, of a continual battle. Discouragement, cynicism, doubt, will allow this continual bombardment of energy to do damage through its natural ten-

THE RELIGION OF PHYSICS

dency to clean out the stalls of the old to make way for the new through suffering (or the energy called Siva in Hinduism —or Trump by Republicans). So-called enlightenment upon the unprepared personality results in disaster.

But remember this: By building the antahkarana a person is speeding (spiritual) evolution, which is jumping beyond the normal rate of the spiral process shown in the chapter on Purpose. This speeds our own karma waiting in the wings, and this miserable (but cued to teach us) "stuff" comes at a much higher frequency. It also will occur immediately after a misdeed with no drag time. One must walk a tightrope, ever observant, or else suffer. Not everyone wants to do that—in fact, very few do. That's YOU, if you have made it *this* far.

We can be a bit more scientific about this science of the antahkarana, and say that the building of the antahkarana, the beginning of the so-called Path, Figure (59), starts at the mental unit (MU) and proceeds to the mental permanent atom (MPA). We can define the mental permanent atom (MPA) as that point created by the intersection of the line of time of the monad (our basic unit of individuality) with the energy field present between the soul and the personality.

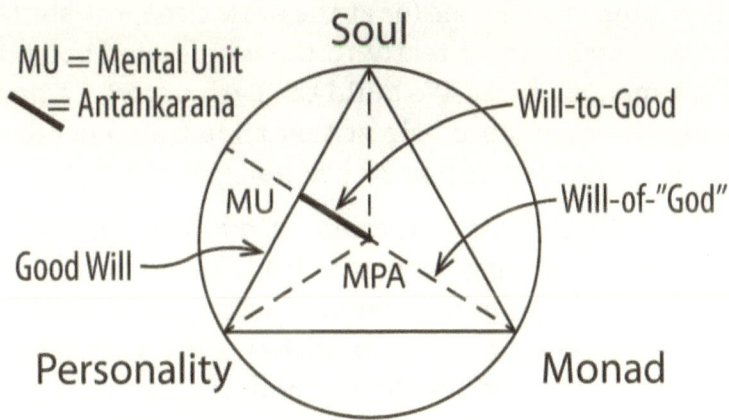

Figure (59) Building the Antahkarana

The mental unit, placed between the soul and the personality, resides on the time passage that connects the mental permanent atom and the monad. The mental unit, in fact, is the intersection of that time passage with the energy interaction between the soul and personality.

In building the antahkarana, we begin to build the Path by use of all the powers we can muster from both the soul and personality, and we forge ahead by use of the Will-to-Good. Esotericism calls this the beginning of Discipleship. We have the capacity to 'reach' the mental permanent atom (MPA) by use of this Path which really is only a line of time (a potential happening, a possible happening) in the time dimension of the monad (the energy representation of unity within us--Oneness). We find that the mental permanent atom resides at the intersection of the lines of time of three separate dimensions of time. Such a situation means a major shift in the manifestation of our being. We find ourselves at the end of the line of human evolution as we know it and the beginning of the line of another evolution, a *new species*. Remember, I talked about a new species in Chapter One?

Meditation can speed the process of soul-personality integration, and thence (or at the same time) the step beyond the soul, and going directly to the wholesaler, but not quite to the manufacturer. We build the Path ourselves, and fortunately, we have more help at that time from energies of our subjective (soul) group and its center of energy.

In our everyday experiences, building the antahkarana means that we must (figuratively) take the first step of the Cross, which is, as you remember, seeing the symbolic (other religions in their esoteric branches use different symbols). That means that we see the symbolic, the spiritual, in everyday life. That does not mean seeing the paranoid, as that is the flip side of this whole process, and a terrible burden. If we can see the symbolic, glory will follow if we also continue to abide by the Will-to-Good. A simultaneous orientation will key the mental permanent atom, and intuitive knowledge will result.

Art

All that we experience has abstraction behind it. Generally, we have no awareness of that, except in the case of art. So, let us take a brief look at art and the abstract, and then move on. Some modern artists attempt to portray the abstract, but all artists deal with the abstract, even if their work is representational. We know that an abstraction points not only to the object but also inwardly to being. Looking at art, say, by Mondrian, we may not know the experience or the object involved, but we can be drawn to the work, feel that in some way it has significance for us, not knowing how, but still sensing that it touches our being.

A piece of abstract art generally has two components: line and color. Line derives from thought, and color from quality. These two factors, plus that of intuition, synthesize to produce the work of abstraction. Through the artist's experience, we have touched our own being, and with it, the being of all. We have a feeling of freedom, which is the basis of faith that we are on the right track.

The visual experience carries us to Being, without effort on our part, except to go to the gallery, of course. We see the work of art as a gift to us from the artist. This is an act of love on the part of the artist, and it remains a unique experience. Most such experiences outside of art require us to exercise our creative wills at some point in the process of synthesis. Such is creative work. Art is a free ride, if we want to take it.

As humans, we are more of a work of art rather than a machine of flesh and blood, a being and not a living computer. We all have being in common, and our work, which is an expression of being, is the art in life—at various levels of expression. Try teaching that to a computer.

Application of Synthesis

Synthesis, the process itself, relates the spiritual and the physical with Being. Being produces Life and is not itself changed in any respect. Synthesis exposes Being.

Life remains immediate and independent of the relativity of the powers of consciousness and evolution. We see two processes working together to produce existence (1) Spirit, the powers regarding the evolution of consciousness, the sphere of relativity, and (2) Being, as it creates Life and ensouls, synthesizing live units of consciousness. We find, therefore, with these two processes working simultaneously—one immediate and the other evolutionary—the same with the immediacy of quantum entanglement and relativity limited by the factor of light. Both the evolutionists and the creationists are correct. The processes work together to produce existence: one *immediate*—Being, and the other *evolutionary*—Spirit. Each requires the other.

The evolutionists and the creationists should, therefore, work together just as the powers they expound work together to produce the universe. The only difference between a so-called evolutionist and a creationist remains the type of energy each is pointing at. They argue, because **they do not realize that they point to different energies, both legitimate energies, both present, but one immediate, and the other evolutionary**. To me, this last statement is the most important of this entire book, and it has significant implications. I could not make the statement at the beginning of the book, because without all these pages and pages of details (and I appreciate that you have gotten this far), the statement would have made no sense.

If we look at any process of creation, we see that the matter created (or existing from a previous universe if you want) can only react or change within the physical parameters set by Being. Being sets the possibilities. Then the process, as it evolves, is one of Being as it is expressed by us (and everything else), which we call Spirit. We seek the spiritual, because of that little Divine Spark within us called being.

Being (as Life) manifests differently depending on the quality of the substance involved. Energy, to us, is essentially energized substance. The goal of evolution appears to be at-one-ment (atonement) which functions as an uploading of all substance such that the spiritual aspect becomes more and more apparent.

In figure (60) life evolves held together by the direct line of Being, but spirals away from the physical to be then drawn more strongly by the energy of the Spirit back to Being. In that process, the levels of consciousness increase, and our permanent atoms, which are our centers of being (also called individuality or monad), move right along with us to anchor the 'new' level of consciousness within us. Energized substance becomes finer, and as I pointed out in the chapter on Healing, we may someday be able to do surgery without cutting.

In our own lives, we strive to particularize the universal through what we perceive as our role in life. We set ourselves upon the so-called Path when we begin to realize our role in conjunction with our subjective group (the Soul). We begin to sense Plan and experience some vision and intuition, and we consciously direct our service to greater effect. People so involved are, esoterically, termed the *New Group of World Servers,*

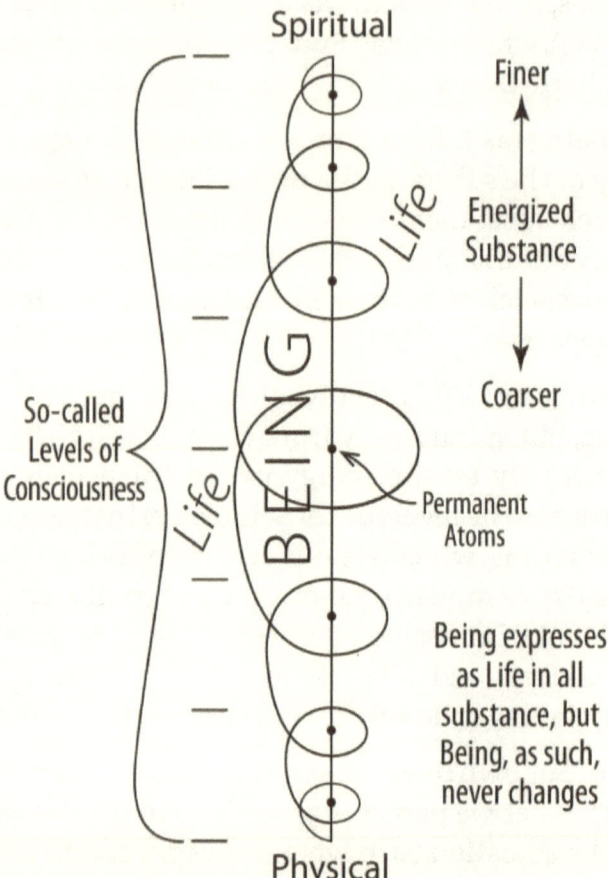

Figure (60) Evolution of a Human

though most such people do not know the terminology or even care to know. They are too busy, serving. Such people recognize one another intuitively, however. It is the manifestation of Being which they recognize: a quiet, knowing pleasure.

Do we have help in these processes? Help is accessing ideas by touching the Spiritual level of existence via the Eternal Now, and ideas then flow through the Antahkarana.

That is help. It is personal to us as it is coming through

our personage.

Those people who wish to accelerate this entire process find themselves within a group called the New Group of World Servers, a subjective grouping of energies held together by **motive**. It's all about motive. But in humans, it is more of *motif*, because we are each a work of art.

Are there people who are more aware of abstractions behind ideas? Yes.

Do we see these people, or people like them, advertising their services? No.

Is there physicality more abstract than us? Yes. Any abstraction has physicality associated with it, but it would not necessarily have to be within our dimension of time. Do these consist of teachers? That is something you will have to find out for yourself.

Regardless, you do have help.

The spiral of existence proceeds into greater and greater abstraction to reveal Oneness. We cue imagination and intuition, and thence the (good) will to discover becomes the courage to be. Thank Paul Tillich for that.

Recognition of Being

In the recognition of being, we know that a trinity of forces is involved: The Universal, The Soul, and The Particular (personality in our case); or Father, Son, and Holy Spirit in Christianity, or whatever, in what-

Figure (61) Heart Integrates a Trinity
of Energies to Expose Being

ever religion(s). We sense the necessary triple alignment, always expressed in service. The most frequent symbol we see that points to this process is heart, Figure (61).

Heart is the alignment of a trinity of forces. We can use that symbol at any level of consciousness, and it exposes Being. When we expose Being, we redeem the earth, in our own way.

So, is God Energy-in-itself? We know that all is energy, and energy is, to us, energized substance (revealing force). Holding the concept that God is energy and staying within the universe (energy always being applied as force) is as far as an atheist can go to a God. Accepting Energy-in-itself includes the universe, and might also include anything beyond the universe, in which case a person (a Deist) can believe in a truly ultimate God. Or, a person can apply the knowledge of energy to her/his own particular philosophy or religion and find benefit.

Pervading all is Divine (or Universal) Selfness or Being, the universal synthesizer that produces Life. Any concept of God includes all of that—the physics of the universe. Beyond that is Energy-in-itself, which is the energy divorced from its application as force. All that is E. Call that step metaphysical, if you want, or call it God, since it is all energy.

THE RELIGION OF PHYSICS

To be, or not to be, *is* the question.

APPENDIX: LIGHT AND SOUND

What follows is esoteric with some speculation and not of interest to most. That is why I have placed it as an appendix.

When something moves, it gives evidence of its movement to surrounding systems. The evidence of the move will consist of a vibratory activity with a constant or variable frequency. Because of the movement, waves of energy transmit thorough the surrounding systems. In our case, we might hear the movement.

We cannot really "know" something by means of these vibratory messages unless we know the character of the surrounding systems and how they interact with these messages. Clicking stones together under water sounds different than clicking them in air. If we know how the surrounding media or systems affect these messages, then we can infer or translate the character of the original reaction.

When we observe light from outer space, we know that light represents the result of some reaction taking place, and this light has transmitted perhaps millions of light years to reach us. Unless we know something of the media through which the light transmits and how that media and the light waves interact, then we cannot accurately know what those light messages mean.

To have evidence, of course, we need an observer. With

the observer, we then have the same problem, because the observer interprets the vibratory messages. Then, we have to depend on the assessment of the observer.

Interpretation

We may, in our culture, react to certain things—say poverty in our midst—with a variety of reactions, not knowing or realizing that our reactions originate from preconditioning present since we were born. We look at poverty with conditioned eyes. When we see people impoverished, then we cannot know what we see unless we know what our own systems do with those transmissions as they propagate through us. If we call ourselves the medium through which the message of poverty transmits, then the medium is the message.

If the governor of a state sets certain policies into effect and her/his only feedback comes through the governor's staff, then the governor's knowledge of reality may find itself far removed from the real effects of the policies. If, however, the governor visits the people affected by the policies, then the consequences of the policies come into better focus—not complete focus unless the governor knows about his/her own preconceptions.

We must deal with the problem of our reactions to any in-coming vibratory messages. Knowing something about ourselves seems a likely answer to overcoming the effects of *illusion*. Each of us has limitations depending upon whom we have known and what we have done. We all have limited ring-pass-nots for our perceptions.

All sound represents communication, equally true with *all* sensory input. We have sensory input at all angles, and each increment of sensation expresses something, for instance, the sounds of language, or the sounds of nature. Each sensation tells us something of the origin of that sensory input.

Do each of our senses operate independently of one another, or do the senses relate in some way? In particular, can we find a relationship between sound and light, and if we find one, does that relationship have any direct importance to us?

Light and Sound Relate

Plan organizes according to light. Light in all its manifestations—from cosmic rays, to invisible and visible light, or to heat, or even to electrical waves—in all of these, light makes things happen. Visible light forms only a part of the electromagnetic spectrum. **When I speak of light in this chapter, I am referring to the entire electromagnetic spectrum.**

Just as the seed represents the ultimate abstraction of the plant, light represents the ultimate abstraction of all creation. Light contains the basic genetic code for creation. With the study of light, we will discover the secrets of the universe.

The entire spectrum of light rains on us every day. We do not have a conception of how all the components of light interact to do what light does. Because of light, only certain things *can* happen, and certain things cannot happen because of light. Light sets the possibilities of creation. Light is Plan. Light affirms creation. Light affirms the world of non-being. Light exists as the only visible representation of being.

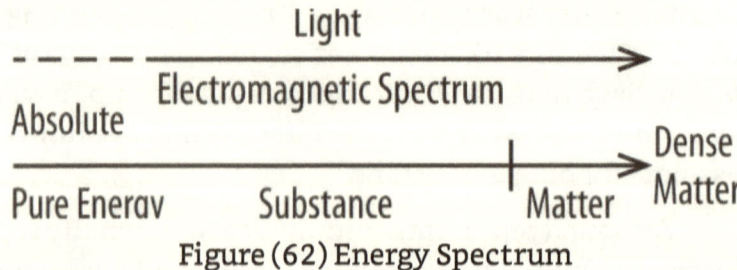

Figure (62) Energy Spectrum

Light cannot do anything without matter. Light must have relation to matter or else light does not exist. We can look at the universe as a spectrum from spirit to matter. We might call this the Energy Spectrum, Figure (62). At the beginning of the spectrum we find the Absolute, that which we could call pure Energy-in-Itself, Pure Being, whatever you want to call it. At the other end of the spectrum we find dense matter. Between the two we find substance and light. We still know little about light, but light contains the coded message. When light combines with matter, then we see both the qualities of light and the qualities of matter.

When we expose the mind to light, we find immediate knowledge, or intuition. We want to access light in our minds and learn from it. Our minds serve as organizing tools to this in-coming data. What we call Plan consists of the encoding of light exposed to mental substance, but mental substance associated with Plan must originate from a greater being or a higher organization of information than a human. We associate this mental substance with the etheric body of the earth—the energy atmosphere of earth.

As humans, we cannot generate light. We can reflect light. We can serve as vehicles for light, but we cannot create light within ourselves. We must seek light through sound. How might that work?

We remember that sound must have matter for transmission. Sound exists on the matter level (mother level), and does not extend into substance, Figure (63). What we identify as the spectrum of light resides only within substance, substance consisting of the media through which light transmits. Again, substance makes up the etheric envelope of the earth.

We therefore have an interface between sound and the electromagnetic (light) spectrum. Once humans began to use the electromagnetic spectrum with which to communicate—radio and television—then we feel in a practical way that sound has connected with light even though we still do not have a theory that connects the sound spectrum with the electromagnetic spectrum.

```
Electromagnetic Spectrum  |  Sound
─────────────────────────────────
        Substance         |  Matter
```

Figure (63) Sound and Light Interface

The Interface Between Light and Sound

The connection between light and sound, proven or not, remains our access to light. Between pure being at one end of the continuum and dense matter at the other, we find all the manifestations of energy. Beyond sound, we find light. Let us examine the interface between sound and light. Across this interface we want to build our bridge to the universal or spiritual level of being. That bridge is the Anatahkarana.

The interface between Being and Non-being we can describe as zero, Figure (64). The zero limit defines the ring-pass-not, or the ring of consciousness. Being affirms non-being. In previous chapters I described any object as a necessary 'hole' in non-being, necessarily filled by whatever material non-being is. To move from non-being to being, we must go through zero. Such a move expands our ring of consciousness, the ring-pass not.

We have associated sound with the material aspect of life, as sound requires a material medium in which to transmit. Without a material media we do not have sound. Light on the other hand has no media yet identified by physics. I am saying that light and sound reside on the same spectrum, even though light associates with the media of substance rather than matter.

We all realize that the entire electromagnetic spectrum falls upon us every day, seemingly in a disorganized manner, much like rain. Yet if we use instruments, we can discriminate the various components of

THE RELIGION OF PHYSICS

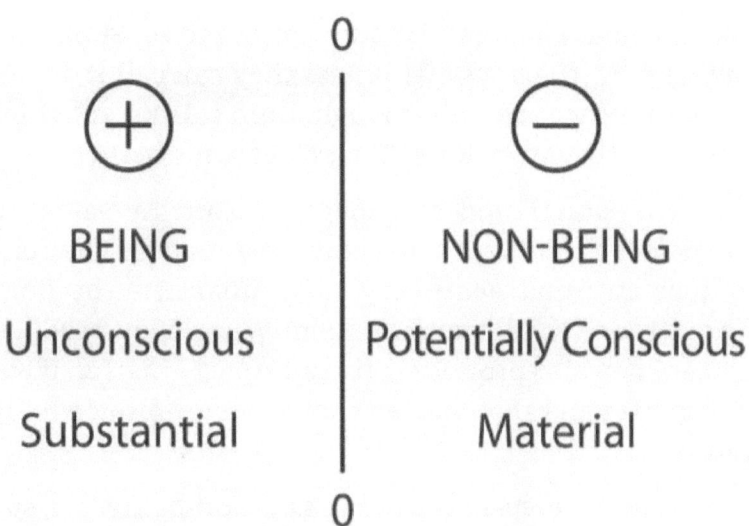

Figure (64) Zero Separates Being and Non-Being

light: cosmic rays, ultraviolet rays, visible rays, whatever. We can easily say that all light contains information of the source of that light, but we have light from billions of different sources hitting us. It all looks like light to us when we look out the window.

If sound could transmit as well as light, then we would have a similar situation, and sounds of the universe would bombard us at all times, and we would have that as a background sound. We would have a non-discriminatory approach to that sound much as we would have to light, saying only that what we hear consists of all sound, just as we refer to light as all light.

One sound interacts with another sound to produce a different sound. Similarly, one color of light reacts with another color of light to produce a third color. We have many similarities between light and sound, but science denies that they exist together on a spectrum only because they have not figured out any medium in which light transmits.

Substance and Anti-matter

Light associates with substance, and substance remains a mystery to the physicist unless they can call it a name and prove its existence. Fair enough. **I am calling substance the same as anti-matter.** Bong. That rings some bells.

We cannot find anti-matter in our live-a-day world. Yet matter cannot exist without anti-matter. Physicists can produce anti-matter in very small quantities by bombarding matter with high energy light. When they produce anti-matter, they also produce its counterpart--matter. If we combine matter and anti-matter, then what is produced? Light, of course.

When we speak of being and non-being, we speak of anti-matter and matter. In Figure (64) we saw how being and non-being interface at zero forming the limits of the ring-pass-not for being in that particular system. We can say the same for matter and anti-matter. In Figure (65) matter and anti-matter relate to one another by the barrier of zero.

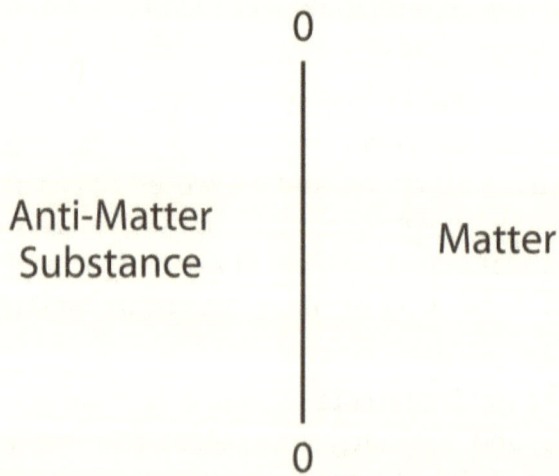

Figure (65) Zero Separates Matter and Anti-Matter

When we expand consciousness, then we expand the ring-pass-not to greater inclusiveness, but we continue to have the ring-pass-not and zero., We always have this zero barrier, as this barrier represents the limits of the ensoulment of the thing or organism. If we want to expand that ring of consciousness, then we begin to create new forms, more inclusive forms. This holds true of humans as well as things. It does not make any difference how expanded in consciousness we find the organism, we can still say that the relation of the realms of being on each side of the barrier of zero remains that of matter and anti-matter.

Plan consists of encoded anti-matter. That is a big statement, but it makes sense, though there's no physical proof. But if true, it then follows that (what are called) Rays make up the broad categories of encoded anti-matter, and within each group of individuals (soul groups) we find only one ray. All rays make up all soul groups. Rays are bundles of energy.

We find ourselves in the process of refining our conscious mental matter to coincide with soul-grade material —embodying our highest aspirations and human qualities. With this matter we attempt to begin to build a bridge to the substance of being. Regardless how pure we construct that material bridge, we still run up against the zero barrier, and to us, it appears as darkness—without light, though we suspect that light exists there, but perhaps in an invisible portion of the spectrum. We cannot prove that it does exist there. We only see blackness. How do we penetrate through? That takes us back to sound.

We can produce sound since sound exists in our material realm. We cannot produce light. We will use a sound of power, propelled by will through the highest grade of mental matter we can construct. We can use the will aspect of the particular Ray involved with our own group to boost energy for that sound of power. We can use that sound of power to penetrate into the zero barrier. What happens than?

Initially we will have an interaction between matter and anti-matter: light results. This light has organization. It originates from Plan (encoded anti-matter). This light rains upon us, and our mental tool of the mind begins to interpret this light in terms of knowledge. We have a sudden realization (or gradual realization). Over time we begin to instill the knowledge encoded and transferred by anti-matter into our energy bodies within our ring-pass-not of existence. We become adept at this, and we have enhanced the transmission lines necessary for direct contact with Plan.

A New Human Species

Once we have consciously broken the zero barrier, and our center of being is beginning to build itself a new temple, then our ring-pass-not expands beyond what we consider the human realm. We have to consider a new species of human, though we cannot call it new in any respect, only new to us. Homo spiritualis is perhaps a good name. Homo spiritualis consciously lives in both the realms of substance and matter —matter and anti-matter—one foot in each. Homo spiritualis, as a vehicle, produces light through service.

Homo spiritualis represents a transition form, a highly specialized form of human to produce service of a particular quality and power. We call that human a disciple. As that human resolves the transition, and the ring-pass-not fully expands into the substantial realm (realm of Spirit), then that human withdraws from the material realm of existence altogether. Yet that human has connections to the Homo sapiens realm, and I know next to nothing about that.

The spectrum of energy contains all the genetic coding necessary for the evolution of the universe from start to finish. Within the bounds of this spectrum, which to us consists of light and sound, all the boundaries and all the possibilities exist; and the universe will proceed according to these possibilities. The ultimate outcome remains determined by the greatest abstractions. A Creator would be the ultimate abstraction, if you accept the existence of one.

The Task Ahead

We now know something of what we want to do. We want to pierce our ring-pass-not, the zero barrier, and access the realm of substance, the arena of ideas. Once we penetrate zero, then we have access to anti-matter, and anti-matter plus matter will give us the light we seek.

An idea can manifest in consciousness only if we set the necessary conditions. We must seek the idea through a meditation or thinking process by building thought substance up to a point that we can go no further, reaching the limit of our thought on the subject. This thought substance must have the highest quality according to our capability, meaning that the *motive* behind the thought processes must contain our highest aspirations. We rarify this mental matter by its reflecting the best in us, and by so doing, we make it possible that we can use this matter as a bridge, however you wish to picture it.

Sound as Projectile

The passageway through this soul quality matter goes by way of the will. When we speak of the Antahkarana, then we speak of a way guided by the passage of Divine (Universal) Will, or we can call it the way of Being. We must exercise Will of a type consistent with the Ray makeup of our soul-grouping. I will again call this the will to discover.

We now have the medium, and we have light on the path in the manner of our projected will. We only lack the projectile to send through this channel which will penetrate the darkness of nothing—zero. In reality, *nothing* does not exist. What appears as nothing to us merely represents zero, and even within zero we will find energy.

We use sound as the projectile, and once we have penetrated the darkness, then we encounter anti-matter, and that anti-matter interacts with the rarefied mental matter that we have collected with our thoughts. Light results and that knowledge answers the question that we had-and much more. Ask and it shall be given. The answer is a matter of physics. I also believe that it *can* be more than that.

I have to say here that the entire process I am describing occurs automatically. I am only trying to describe as best I can with what incomplete knowledge that I have of the process. When a person learns about meditation, it sets up the infrastructure so that changes are possible. I am trying to explain the infrastructure. If I am wrong about the details, it will not make any difference if you proceed according to how it is done, and you do it YOUR way. Process is less important than motive. Motive leads the way.

It might be asked if a pressure gradient exists across the zero barrier, and because of this gradient the energy, anti-matter would extrude or flow out, so to speak, once hit by the projectile of sound given the proper conditions. Light in itself *can* exert pressure. Physicists know, for example, that a flat piece of mental foil floating in space will move if exposed to high energy light. Light exerts pressure. The pressure of light guarantees that the ultimate direction--assuming that all other reactions neutralize each other--the pressure of light will determine the flow of the universe. Thanks for that. A reason to be positive.

Overcoming Inertia

As we begin to access the substantial level of being more times, then we are having more and more flashes of light into our system, and the pressure begins to build within our own ring-pass-not. The pressure of change builds, but the ring-pass-not does not budge. We do not expand our ring of existence or consciousness gradually. The system does not work that way.

All systems have built-in inertia to keep them as systems. We call this sustaining energy (Hindu), or ensouling energy, or energy of Son as in Christianity. As the pressure of change builds, the resistance holds for a while, but sooner or later we will break the bounds of the inertia and have a sudden realization, a quantum leap. In medicine we call this the threshold phenomenon, a common fact in biological systems.

The quantum leaps of change always occur within pre-arranged limits since we have a system made up of certain energy configurations that limit the possible forms. The system has a Plan. Important to remember that one's ring-pass-not limits the quality of the light or information coming it. There's a limit. Hence, we know why the Raincloud of Knowable Things has that limiting term "Knowable."

First Initiation

These quantum leaps are called initiations. The first two are minor, but the third is major. We know when we have them and they make big differences in our lives, not so much in what goes on in our lives, but more in how we see the process of living. We then make adjustments, always for the better. We certainly do not see harmful effects from the process, though we might make fools of ourselves from time to time.

The first initiation we can call the beginning of the growth process. At this initiation, the person becomes real. We can also call this the resolution of adolescence, though many people do not go through this change until they are in their 20's, or near age thirty (as did I). Prolonged education seems to extend adolescence. Most people do not go through this change at all. We can call this initiation one of **birth,** since from that time on, significant personal growth occurs.

At the time of this initiation, the person begins to make contact with soul energies in such a way that he/she takes on a wider purpose in the affairs of life. One does not always identify precisely the character of this purpose, but the person becomes a seeker, and in this seeking process we see growth of the personality in relation to soul qualities—closer to the interlude (the Eternal Now—Zero Time, etc.). My second book, *Personality and the Soul*, examines this process through personality spectrum, using sixteen community women as examples.

Such an initiatory experience begins the connection between the lower and the higher (the personality and the soul) and the person emerges from the unreal to the real. The line of energy connecting the soul and the personality becomes an enhanced connection.

The mechanism of growth during the period after the initiation and up to the next initiation consists of examining one's activities in the light of the soul, rather, examining one's values in relation to soul qualities or energies and then bringing these souls qualities into one's everyday life—the light of the soul. That is called the beginning of the integration of the soul and the personality, also called a process of personal growth

If I had to characterize the first initiation, I would do so by calling it a time of giving up. We give up the fantasy that we have *all* the power, that we have omnipotence (I think, nowadays, Google accentuates this fantasy). We come to the realization that we dangle from the end of string held by another Hand. Once we give up, then we become real. Taking responsibility for one's actions follows, and we build lives according to a higher standard that becomes second nature.

We find ourselves born into a new phase of the life process, and the type of Will we use to push this process forward we call Good Will. Creativity during this phase tends to ground itself in reality, a reality we are discovering in ourselves as well in others and in nature. We learn that when we truly listen, others speak profoundly.

An initiation represents only a beginning within a new realm. At the time of the initiation we have a realization of what lies ahead. This realization gives us enthusiasm, and we want to forge ahead. When we refer to the initiation itself, then we refer to this glimpse ahead, which is a more inclusive glimpse of reality than we had ever experienced before. The initiatory process continues all the way through until the next initiation.

Second Initiation

The second initiation deals with submergence in the universal solvent that appears at first to us to consist of feeling, and for a time, we battle with the control of our emotions. Yet, this solvent does not consist just of emotion. We are dealing with something that makes us all part of one system, oneness basically. This initiation occurs with just the first inklings of the idea level, also called by esotericists the Buddhic level of existence. Within the Buddhic level we find the universal aspect of emotion, and though we have had only a glimpse of that world, we now know what we want. We call this **Baptism**, similar to religion.

If I had to characterize this second initiation, I would call the process one of purification, as we further pull our lives away from the material aspects and emotions related to those endeavors, and we begin to focus outside of ourselves. Even with the first exposure of this initiation, we have not yet controlled our self-centered emotional lives to any great degree.

The wider realization of the initial experience will propel one to thinking that he/she has become a prophet (profit?) of some sort, and that there is something great to say to nearly everyone. Such is the fog produced when the water of emotion exposes itself to the fire of the mind. Indeed, this entire initiatory experience consists of the higher mind beginning to take control of the personality. Did this book, and my previous books, represent this fog? I hope not.

I do know that we can dispel the fog as we continue to learn to reach beyond ourselves and our glamours (illusions from the emotional level of existence). We cannot suppress glamours. We dissolve them either from the pains of learning

—and we will have real pain and suffering during that period—or we dissolve them as we reach toward the more potent solvent of the Buddhic level of existence, the universal aspect of sensitivity. Our only tool remains our minds, and we grope in the dark, somehow sensing where we want to go, but only seeing the light of the soul on occasion. We have ups and downs, but continue to seek.

Whatever we might have been struggling with in our ring-pass-not at that level of existence prior to the initiation, then becomes instinctual for us. We will instinctually act according our last completed initiatory phase of experience. When we enter into a new phase, then we enter a so-called hall of learning that has an expanded ring-pass-not, but we enter in ignorance. Eventually our learning turns to knowledge.

Third Initiation

The third initiation is the first major initiation, and it requires conscious effort as opposed to the first two initiations which occur spontaneously. After this third initiation we find the human now called a disciple—a highly specialized human who serves. When we speak of using words of power, then we are speaking of this third initiation.

Until the third initiation, the method of bringing existing soul-grade mental matter (knowledge of the soul) into waking consciousness consisted of meditation in which we consciously brought light through the lower bodies to the waiting brain, and then waited for realization. The process I am *now* speaking of involves an intrusive invocation—the sound of power—within the interlude between sound and light.

The third initiation must have conscious activity. We need to learn the process whether by looking within ourselves, or else have guidance. We can find many guides in life, and we will know when we have found what we need. What works for some, does not work for others. Ultimately, however, we are alone, and we must do it ourselves.

We seek to become adept at the process called contemplation. All great thinkers use contemplation. Some have learned it. Others have seemingly come upon it naturally. Contemplation consists of the highest level of the science of meditation (to us).

In contemplation we become one with the object of contemplation, which means that we have created something through the action of matter and anti-matter, and what we have created actually represents an extension of ourselves. We remember that we are reaching toward the substantial level of Being by using the most purified mental matter that we have. We are what we think, and magically the highest extension of ourselves suddenly becomes something so much more, and we open our eyes to a more inclusive reality. Beats the movies.

Progress

With all three of these initiations we work toward connecting the spectrum of light with that of sound. The level we seek, the Buddhic level of existence, consists of the etheric vehicle of this earth. On a world-wide scale, we are now entering a phase with the advent of telecommunications, especially the internet. Regardless, humanity as a race still resides far from the third initiation.

Individually we will use sound sent by way of a path that we call the Antahkarana, and this invocative sound will evoke a light response from the substantial level of Being. When a sufficient amount of the substantial (etheric substance) has integrated into ourselves, then we each will have prepared for the third initiation. This process works within religion as well. The institutionalization of religion, however, hampers that progress.

What is a Sound of Power? I must defer to Simon and

Garfunkel: The Sounds of Silence. Yes, there are sounds . . . Omm, and then there are the sounds of silence. In each, motive is paramount, and each works. In reaching into the interface between sound (matter--our world) and light (the representation of Spirit), and entering Zero, the realm of Being (the world of Ideas), then revelation occurs. One cannot help but believe in one's own revelations. But you will find what Will Durant found: ". . . for in philosophy, all truth is old, and only error is original."

Going it Alone

We are nearly at the end. Some of you have had confirmation of ideas and inclinations that you have had for a long time. If you have not experienced any of what I have written about, then I don't know how you have been able to read this far. If you have, you are likely angry as hell, and you wonder how any such heretic as myself can say such outlandish things. If you are a scientist, and you have experienced some of this or little of this, then you call on me to prove what I say. What is the truth? What is pure speculation? Fair questions.

Science proves. But you cannot prove the future. When a person accesses an idea from the idea level, then these ideas refer to the future, because these ideas are yet to fully express. Ideas are the blueprint, and they show where we have been, and they indicate where we might go. The details of how we will go there, and the technology involved, will change; but the ideas stay the same.

Science must prove. Over the years a certain body of knowledge becomes codified, and we might read these accepted truths in an encyclopedia. These become part of our cultural heritage, and they might also serve as blinders for truths that lie ahead—especially regarding completely new paradigms. Generally, new paradigms are happened upon quite by accident. How could that be otherwise, if the ideas are new?

When dealing in religion and philosophy, we cannot demand proof. Most people feel satisfied that there exist holy books that contain the truth. These are books accepted by millions of people over the millennia, and these books, like the encyclopedias in the secular world, serve as a cultural foundation. Holy books do not change as we learn more. Holy books are supposed to contain everlasting truth.

We still do not know what is true. Someone saying that a holy book is God's word does not make it the truth. We are left, ultimately, alone; and that is perhaps why people look for intermediaries, such as a Christ or Mohammed; and in that way, they do not feel so alone. At least there is a person they can relate to. But the fact is, we *are* alone, because we are the ones who must make our own decisions. Belonging to a congregation of fellow believers helps in some way to give the courage of beliefs, but congregations can err, as history has shown. Just because a group, no matter the size, affirms a belief system does not make it the truth.

Again, the bottom line is that we are alone. We, alone, must decide. We have no tools but those that reside within us. Esoterism gives us some external tools, as that is what esotericism is; but we have to use the tools ourselves. I can only speak from experience, and you must take that for what it is worth (to you). I have the ability to access ideas. Ideas come to me in conjunction with my studies and thought, and I write them down. Sometimes they come in a rain, as I have described. This sort of experience has been documented many times by others. Does that mean these ideas point to truth? I have to say that some of what I have said in this book is quite fantastic. I cite the chapter on Will and Space in which I show matter decreating and creating to give movement, and my equating anti-matter and substance. It all makes sense, and it fits nicely. But that does not make it true. As a physician I am a scientist, and I carry a considerable amount of skepticism. When material comes to me, I know much of it *is* true, but some of it, to be exact, is way beyond me. I have no inkling of its truth, even though it makes sense to me. We'll see, eventually.

I believe in what works. What might work within a very narrow community might not work within a wider community. We each must pick out what works for us and be sure we do no harm to others in the process. That last statement is most important.

As a physician, I help others to figure out what works best for them. What works for them, does not necessarily work for me. We are all different, and each must make his/her own decisions given the situation. And we each are 100% responsible for our decisions.

We have to live our decisions. It is in living-out our decisions that we discover truth. We receive confirmation or non-confirmation of the ideas that bolster our actions by our living our decisions (sort of like the results of an experiment). Each idea has many ramifications within our daily lives, even some of the very far-out things that I have talked about. At some time or another we will receive confirmation of the idea, and we might remark that we have come full circle (the spiral). Each of us have had that experience. That experience confirms an idea (or a so-called truth) for us. That confirmation gives us the courage to continue—the courage to be. Without that confirmation, we would be suffering from depression.

Esotericism is only a tool, and you should be able to use the tools within you own belief system and not change the system in the process. Likely your system contains considerable truth. But as I said in the first chapter, exclusivity is the only thing that throws a wrench in the works. You cannot have exclusivity and seek truth.

What is truth? You can take accepted standards and lead a full and happy life within these standards. You still have to make difficult decisions to follow those standards, and you must put up with the guilt if you do not adhere to the standards.

The other choice is to look within yourself for the truth and seek the truth therein. That's the big choice. That is the choice of esotericism.

We cannot trust people who claim to have found the truth within themselves and then suggest or insist that we should follow them. There are too many clever psychopaths in the world who sound very sincere and have convinced even themselves that they know the truth. When they are *that* convinced, they can easily find other people who will believe in them, just because of that self-assuredness. Dangerous business.

Before you make the decision to go it alone, you must have some soul infusion of your personality. That means you must be real. The only hooker here is that some very disturbed people are still very real. This is the problem of psychosis. If you lead a productive life, and you are able to have a sustaining family life, and you can find happiness in what you are doing, then you are not psychotic. A person unfortunate enough to suffer psychosis has a great deal of difficulty maintaining a stable life. They need help on several fronts. Hopefully, this country will get smart enough to provide help to these people and continue that help as needed.

So, are you ready to go it alone? Well, actually you already have your subjective group with you. This is your soul grouping, and that's a real group of living people. As part of this energy-related group of people, you are actually working toward a positive goal. You may not know the goal, and you may play a small part in it, but there are other people who are doing the other parts—it's a group thing. You will have inklings about this, sometimes, strong inklings. If you are lucky to meet a member of your own (energy or soul) group, there will be immediate connection, like no other you have experienced. Yet, that experience might be very brief.

The more soul infusion within your personality, the more you can consciously access that group as a reality. The group has a task, and your job is to become a better vehicle for that task. Such is the definition of service. The subjective group replaces a congregation within religion. Nobody ever said that groups were bad. If you can do something individually, then you can do it even better in a group. That only makes sense, because there is inherently more power or energy in a group.

The group's task is to hone the powers of creativity with the goal of service. Ideas come from the idea level or Plan. The conscious integration within the soul has its value not only as a vehicle for service, but also protection. The energy coming from Plan is high voltage stuff, especially within the Antahkarana—the self-constructed direct pathway. Without good integration with the soul, the energy of ideas can blow you away. That is why some very real people have problems with what we call psychosis. Such is why many creative people die quite early in life—especially mystics. The voltages brought in will boost all systems, even the defective ones. The defective ones may bring an early demise.

This adventure and all the possible dangers to be faced is not everyone's cup of tea. For some, however, it is the only choice. Those people will accept nothing else. Some have called these people The New Group of World Servers. All that means is that there exists a group of people throughout the world who consciously seek service through doing their own thing. They are part of that group, however, because they seek service though the guidance of their group within. They have a natural tendency to do that, and they don't consider it anything but what they want to do. That's fine.

For those of you who want to firmly stay within a formal religious grouping, then esoteric thought merely gives you another tool to grow within that religion. The tool helps you to evaluate. Not everything within any religion will ring true to you. You need to have faith in yourself in that respect. All religions have their esoteric aspect and some even have formal esoteric branches. They don't talk about that, but if you ask around, you will find some pretty amazing things.

Some of you may want to go ahead with further esoteric studies. You aspire to something. I hope that is different than ambition. You are what you are. You are no more than that. You can better develop yourself, and you can better realize your own potentials, but you cannot POOF! become a great spiritual leader or a guru to others. If you want to become a leader like that, then go ahead and enter esoteric studies with that ambition, but you will cut yourself to threads.

If, with all these warnings, you still want to enter the study of esoteric matters, then you would do well to have some direction at first, perhaps for years, perhaps for the rest of your life (I was offered that). I would recommend an esoteric school such as the School for Esoteric Studies in Ashville, NC, or the Arcane School in New York, or perhaps *The Rainbow Bridge* by Two Disciples, The Triune Foundation, 1981. I would be wary of any esoteric school that might offer certifications or any degrees, or charge high tuitions. This is not something you can get a degree in, though some might counter by saying that religions certify their ministers of the faith. Why can't esoteric studies certify their teachers? I suppose it's possible, but once you begin to institutionalize anything, then you find built-in blinders and competition with hierarchies. None of that jives with esoteric thought.

What you have read in this book is MY life. If you can find help in your own life reading about mine, then I have been successful. I am not adept in esoteric studies, and I am not a leader of anything. I am only one, as are you. Welcome to the group.

Comments: pgroce@roadrunner.com.

Please consider commenting on the book at the Amazon site. Thank you.

CREDITS

Portions on Bargaining and Argument in the chapter on Light were excerpted from the author's previous book, *When Mirrors Become Windows,* Northwoods Press, 2001. Copyright Philip C. Groce

Portions of the chapter on Personality were abridged from the author's article "Personality Spectrum" in *The Journal of Esoteric Psychology,* University of the Seven Rays Publishing House, fall/winter, 1998-9. Copyright Philip C. Groce.

Portions of the chapter on Purpose and also Will and Space were abridged from the author's article, "Esoteric Physics," in *Beacon,* Lucis Trust, New York, Volume XVII Number 1, Jan-Feb, 1999. Copyright Philip C. Groce.

Some material pertaining to ideas was abridged from the author's article, "Ideas" in *Beacon,* Lucis Trust, New York, Volume LVII Number 5, September/October, 1997. Copyright Philip C. Groce.

An edited section on the process of Bargaining in the chapter entitled Light was published in *The Journal of the Maine Medical Association* entitled, "Marital Counseling for the Family Practitioner," by Philip C. Groce, MD, 1974. Copyright Philip C. Groce